OXFORD MATHEMATICAL MONOGRAPHS

Editors

I. G. MACDONALD   H. MCKEAN   R. PENROSE

OXFORD MATHEMATICAL MONOGRAPHS

A. Belleni-Morante, *Applied Semigroups and Evolution Equations*

I. G. Macdonald, *Symmetric Functions and Hall Polynomials*

J. W. P. Hirschfeld, *Projective Geometries over Finite Fields*

A. M. Arthurs, *Complementary Variational Principles (Second Edition)*

N. M. H. Woodhouse, *Geometric Quantization*

P. L. Bhatnagar, *Nonlinear Waves in One-Dimensional Dispersive Systems*

N. Aronszajn, T. M. Creese, and L. J. Lipkin, *Polyharmonic Functions*

J. A. Goldstein, *Semigroups of Linear Operators and Applications*

M. Rosenblum and J. Rovnyak, *Hardy Classes and Operator Theory*

Hardy Classes and Operator Theory

MARVIN ROSENBLUM

AND

JAMES ROVNYAK

OXFORD UNIVERSITY PRESS · NEW YORK
CLARENDON PRESS · OXFORD
1985

Oxford University Press

Oxford   London   New York   Toronto
Delhi   Bombay   Calcutta   Madras   Karachi
Kuala Lumpur   Singapore   Hong Kong   Tokyo
Nairobi   Dar es Salaam   Cape Town
Melbourne   Auckland

and associated companies in
Beirut   Berlin   Ibadan   Mexico City   Nicosia

Library of Congress Cataloging in Publication Data
Rosenblum, Marvin.
  Hardy classes and operator theory.

  (Oxford mathematical monographs)
  Bibliography: p.
  Includes indexes.
  1. Hardy classes.   2. Operator theory.   I. Rovnyak,
James.   II. Title.   III. Series.
QA331.R725   1985   515   84-27262
ISBN 0-19-503591-7

Printing (last digit):   9 8 7 6 5 4 3 2 1

Printed in the United States of America

Preface

This book is intended for students and researchers in mathematics, engineering, and the physical sciences who are interested in the interaction of function theory and operator theory; we assume that the reader is familiar with the rudiments of both fields. We are specifically concerned with the theory of shift operators, Toeplitz operators, and Hardy classes of vector and operator valued functions. Examples and applications are drawn from the recent literature and, in particular, the authors' collaborative work.

Although there is some interdependence among chapters, the reader is encouraged to proceed directly to the topic of greatest interest, using what comes before for background and reference purposes. For example, Chapter 2 depends critically on only one theorem from Chapter 1, and that theorem can be best appreciated by first seeing its applications.

Chapter 1 develops the general theory of shift operators on a Hilbert space. It includes the Wold decomposition, a universal model for linear operators, the Beurling-Lax theorem, and the lifting theorem. The examples and addenda to Chapter 1 show some of the manifold connections between shift operators and classical analysis.

In Chapter 2 we use the lifting theorem to give a unified treatment of interpolation theorems of the Pick-Nevanlinna and Loewner types. In particular, we derive general theorems on the interpolation of boundary data. The central results here are the generalized Loewner theorem and its consequences. In the examples and addenda to Chapter 2 we sketch the closely related theory of monotone operator functions.

Chapter 3, like Chapter 1, is operator theoretic in nature. It is an account of the factorization properties of Toeplitz operators. This includes both the factorization of nonnegative Toeplitz operators and the factorization of analytic Toeplitz operators. The factorization theory serves several purposes. One is that it is the basis for the factorization properties of operator valued functions given in Chapters 5 and 6. Another is that it is preparation for the spectral theory of concrete Toeplitz operators. In the examples and addenda to Chapter 3 we outline the explicit diagonalization of any self-adjoint Toeplitz operator, including proofs of the main theorems. We also sketch the principal examples of Toeplitz operators, among which are the Wiener-Hopf operators.

Chapters 4 and 5 develop the theory of vector and operator valued holomorphic functions of bounded type on a disk or half-plane and the theory of operator valued inner and outer functions.

In Chapter 6 we present our view of the factorization problem for nonnegative operator valued functions on a circle or line. Because of applications in prediction theory, much has been written on this problem. Our treatment is different in that it is based on the notion of a pseudomeromorphic function. We show that such functions are automatically factorable. As a consequence we obtain generalizations of the classical factorization theorems of Fejér-Riesz (factorization of polynomials) and Ahiezer (factorization of entire functions). In the course of the development we prove an operator extension of Krein's theorem which characterizes the class of entire functions that are of bounded type separately on the upper and lower half-planes. The chapter also includes the operator generalization of Szegö's theorem, which is one of the cornerstones of prediction theory.

The Appendix is a summary of background material on function theory that is needed at various places in the text. At some future time we hope to expand the notions treated in the Appendix more fully and to publish it in a different form.

Even a casual scrutiny of the literature is a humbling experience. Our literature notes and bibliography attempt to trace historical development, to credit certain specific contributions, and to indicate related directions of research. We gratefully acknowledge J. M. Anderson, R. B. Burckel, and M. Heins for generously sharing their knowledge of the literature.

We thank Gian-Carlo Rota for first suggesting the possibility of writing a book. Support from the National Science Foundation is gratefully acknowledged. We have lectured on the subject of this book in seminars and courses at the University of Virginia, and this contact with students has been of great value. James Rovnyak acknowledges support from the Alexander von Humboldt Foundation for the year 1979–80. He gives special thanks to Professor P. L. Butzer of the Rheinisch-Westfälische Technische Hochschule Aachen and Professor M. Wolff of the Universität Tübingen for arranging extended visits to their institutions. During these visits much of the material in this monograph was presented in lectures under the title "Ausgewählte Themen aus der Funktionentheorie und Operatortheorie." We thank Marie Brown and Joyce Stevens for expert typing and assistance in the preparation of the manuscript. Not least, the authors thank the editors of Oxford University Press for their helpful suggestions in bringing the manuscript to its present form and for including this volume in their series of mathematical monographs.

We express our deepest gratitude to our wives, Betsy and Virginia, and to our children, Dore, Mendel, Jessie, Rebecca, and Sarah Rosenblum, and Steven and David Rovnyak, for their support during the writing of this book.

Charlottesville, Virginia   M. R.
October 1984   J. R.

Contents

Chapter 4.   Nevanlinna and Hardy Classes of Vector and Operator Valued Functions   74

Chapter 5.   Operator Valued Inner and Outer Functions   94

Chapter 6.   Factorization of Nonnegative Operator Valued Functions   109

Appendix. Topics in Function Theory   128

Notation

1. *Unit disk*

$D = \{z: |z| < 1\}$; $\Gamma = \partial D$; σ is normalized Lebesgue measure on Γ

$$P(z, e^{it}) = \frac{1 - |z|^2}{|e^{it} - z|^2}$$ Poisson kernel

$\left.\begin{array}{l} H^p(D),\ N(D),\ N^+(D) \\ H^p(\Gamma),\ N(\Gamma),\ N^+(\Gamma) \end{array}\right\}$ Hardy and Nevanlinna classes (Duren [1970], Hoffman [1962])

2. *Upper half-plane*

$\Pi = \{z: \operatorname{Im} z > 0\}$; $R = (-\infty, \infty)$; dx implies Lebesgue measure

$$\frac{y}{\pi} \frac{1}{(t - x)^2 + y^2}$$ Poisson kernel

$\left.\begin{array}{l} H^p(\Pi),\ N(\Pi),\ N^+(\Pi) \\ H^p(R),\ N(R),\ N^+(R) \end{array}\right\}$ Hardy and Nevanlinna classes (Duren [1970], Hoffman [1962])

3. *Vector and operator valued functions*

$\mathscr{C}$: separable Hilbert space; norm and inner product $|\cdot|_{\mathscr{C}}$, $\langle \cdot, \cdot \rangle_{\mathscr{C}}$
$\mathscr{B}(\mathscr{C})$: bounded linear operators on $\mathscr{C}$; norm $|\cdot|_{\mathscr{B}(\mathscr{C})}$
X: Banach space (often $X = \mathscr{C}$ or $\mathscr{B}(\mathscr{C})$); norm $|\cdot|_X$

$\left.\begin{array}{l} H^p_X(D),\ N_X(D),\ N^+_X(D) \\ H^p_X(\Gamma),\ \text{etc.} \\ H^p_X(\Pi)\ \text{and}\ H^p_X(R),\ \text{etc.} \\ \mathscr{H}^p_X(\Pi)\ \text{and}\ \mathscr{H}^p_X(R),\ \text{etc.} \end{array}\right\}$ Hardy and Nevanlinna classes of X-valued functions (see Section 1.15 and Chapter 4)

$\mathscr{H}_{\phi, X}(\Omega)$ Hardy-Orlicz class (see Section 4.2)

$L^p_X(\mu)$ Lebesgue space (see Section 4.5)

4. *Miscellaneous notation*

$\mathcal{M}(u,v)$	page 113
$\mathscr{P}$	page 34
$\left.\begin{array}{l} T(w), T(W) \\ T(a), T(A) \end{array}\right\}$	Toeplitz operators on Hardy classes; see pages 47, 63, 94, 110
$\langle\cdot,\cdot\rangle$	inner product on a Hilbert space
$\langle\cdot,\cdot\rangle_2$	inner product on a Hardy or Lebesgue space when $p = 2$
$\|\cdot\|_p$	p-norm on a Hardy or Lebesgue space

Hardy Classes and Operator Theory

I am certain of nothing but of the holiness of the Heart's affections and the truth of Imagination—What the imagination seizes as Beauty must be truth—whether it existed before or not—

John Keats
Letter to Benjamin Bailey
22 November 1817

Chapter 1

Shift Operators

1.1 Introduction

All underlying spaces in this chapter are assumed to be Hilbert spaces. We write $\mathscr{B}(\mathscr{H})$ for the set of bounded linear operators on $\mathscr{H}$ and $\mathscr{B}(\mathscr{H},\mathscr{K})$ for the set of bounded linear operators on $\mathscr{H}$ to $\mathscr{K}$. Triangular brackets $\langle \cdot, \cdot \rangle$ denote an inner product.

DEFINITION. *An operator S in $\mathscr{B}(\mathscr{H})$ is a* shift operator *if S is an isometry and $S^{*n} \to 0$ strongly, that is, $\|S^{*n}f\| \to 0$ for all f in $\mathscr{H}$.*

It is convenient to present the general theory of shift operators as a chapter in pure operator theory. The central structure theorem is the Wold decomposition, which shows, in particular, that a shift operator is determined up to unitary equivalence by its multiplicity (see Sections 1.3 and 1.4). Operators that commute with a shift operator play a special role in both theory and applications. The study of such operators is begun in Sections 1.6–1.10 and continued in Chapter 3. Other key ideas discussed in the chapter are (1) a universal model 1.5 for linear operators on a Hilbert space, (2) the Beurling-Lax theorem 1.12, which characterizes the invariant subspaces of a shift operator, (3) the lifting theorem 1.13 and 1.14, and (4) a concrete realization 1.15 for an arbitrary shift operator.

At the same time, the study of shift operators should not be separated from the study of examples. The operator multiplication by z on $H^2(D)$, defined by $S: f(z) \to zf(z)$ for all $f(z)$ in $H^2(D)$, is a shift operator with adjoint $S^*: f(z) \to [f(z) - f(0)]/z$. For any Hilbert space $\mathscr{C}$, the operator $S: (c_0, c_1, c_2, \ldots) \to (0, c_0, c_1, \ldots)$ on $l^2_{\mathscr{C}} = \mathscr{C} \oplus \mathscr{C} \oplus \mathscr{C} \oplus \cdots$ is a shift operator. Its adjoint is $S^*: (c_0, c_1, c_2, \ldots) \to (c_1, c_2, c_3, \ldots)$. These examples are sufficient for illustrating the results in the chapter. Additional examples of shift operators are given in the Examples and Addenda at the end of the chapter.

1.2 Partial Isometries

An operator W in $\mathscr{B}(\mathscr{H},\mathscr{K})$ is a *partial isometry* if W is isometric on the orthogonal complement of its kernel. In this case, we call $M = (\ker W)^{\perp}$ the *initial space* and $N = WM$ the *final space* of W.

THEOREM. *An operator W in $\mathscr{B}(\mathscr{H},\mathscr{K})$ is a partial isometry if and only if W^*W is a projection operator. In this case, W^*W is the projection of $\mathscr{H}$ on the initial space of W.*

Proof. Let W be a partial isometry with initial space M. Then $\langle Wf, Wg \rangle = \langle f,g \rangle$ for all $f, g \in M$. The addition of a vector in $M^{\perp} = \ker W$ to g does not change this identity, so $\langle W^*Wf,g \rangle = \langle f,g \rangle$ for all $f \in M$ and all $g \in \mathscr{H}$. Therefore $W^*Wf = f$ if $f \in M$. Since $W^*Wf = 0$ if $f \in M^\perp$, W^*W is the projection of $\mathscr{H}$ on M. For the other direction we reverse these steps.   ∎

COROLLARY 1.   If $W \in \mathscr{B}(\mathscr{H},\mathscr{K})$ is a partial isometry with initial space M and final space N, then $W^* \in \mathscr{B}(\mathscr{K},\mathscr{H})$ is a partial isometry with initial space N and final space M. In particular, WW^* is the projection of $\mathscr{K}$ on the final space of W.

Proof.   If W is a partial isometry with initial space M and final space N, then $W^*(Wf) = f$ for every f in M and $\ker W^* = N^\perp$. Therefore W^* is a partial isometry with initial space $WM = N$ and final space M.   ∎

COROLLARY 2.   If $W \in \mathscr{B}(\mathscr{H},\mathscr{K})$ is a partial isometry and g is any vector in $\mathscr{K}$, then the infimum of $\|g - Wf\|$ over all f in $\mathscr{H}$ is attained with $f = W^*g$.

1.3   Wold Decomposition

Let $V \in \mathscr{B}(\mathscr{G})$ be an isometry.

(i) There is a unique decomposition

$$\mathscr{G} = \mathscr{H} \oplus \mathscr{L} \tag{1-1}$$

such that $\mathscr{H}$ and $\mathscr{L}$ are reducing subspaces for V, $S = V|\mathscr{H}$ is a shift operator on $\mathscr{H}$, and $U = V|\mathscr{L}$ is unitary on $\mathscr{L}$.

(ii) Define $\mathscr{K} = \mathscr{G} \ominus V\mathscr{G}$. Then $\{V^j\mathscr{K}\}_0^\infty$ is an orthogonal family of subspaces of $\mathscr{G}$ satisfying

$$\mathscr{H} = \sum_0^\infty \oplus\, V^j\mathscr{K} = \{g : g \in \mathscr{G} \quad and \quad \|V^{*n}g\| \to 0\}, \tag{1-2}$$

$$\mathscr{L} = \left(\sum_0^\infty \oplus\, V^j\mathscr{K} \right)^\perp = \bigcap_0^\infty V^j\mathscr{G}. \tag{1-3}$$

We call S and U the *shift* and *unitary parts* of V, respectively.

The above statement of the Wold decomposition emphasizes spatial structure. Substantially the same information is contained in the following.

Operator Version of the Wold Decomposition.   Let $V \in \mathscr{B}(\mathscr{G})$ be an isometry. Then:

(i) $P_0 = I - VV^*$ is the projection of $\mathscr{G}$ on $\mathscr{G} \ominus V\mathscr{G}$;
(ii) as $n \to \infty$, V^nV^{*n} converges strongly to a projection operator P;
(iii) $P\mathscr{G} = \bigcap_0^\infty V^j\mathscr{G}$;
(iv) $\sum_0^\infty V^jP_0V^{*j}$ converges strongly to $Q = I - P$;
(v) $Q\mathscr{G} = \{g \in \mathscr{G} : \lim_{n \to \infty} \|V^{*n}g\| = 0\}$;
(vi) $Q\mathscr{G}$ and $P\mathscr{G}$ reduce V;
(vii) $V|P\mathscr{G}$ is unitary;

(viii) $V|Q\mathcal{G}$ *is a shift operator*;

(ix) $I = P + \sum_0^\infty V^j P_0 V^{*j}$,

$\mathcal{G} = P\mathcal{G} \oplus \sum_0^\infty \oplus V^j P_0 \mathcal{G}$.

Either version can be proved directly or deduced from the other. We prove the operator version and leave the rest to the reader.

Proof of operator version. By Section 1.2, VV^* is the projection of $\mathcal{G}$ on $V\mathcal{G}$, so $P_0 = I - VV^*$ is the projection of $\mathcal{G}$ on $\mathcal{G} \ominus V\mathcal{G}$. This proves (i).

For any $n = 0,1,2,\ldots$, $V^n V^{*n}$ is the projection of $\mathcal{G}$ on $V^n\mathcal{G}$, and

$$V^{n+1} V^{*n+1} = V^n (VV^*) V^{*n} \leq V^n V^{*n}.$$

Therefore (ii) and (iii) follow. The identity

$$\sum_0^n V^j P_0 V^{*j} = \sum_0^n V^j (I - VV^*) V^{*j} = I - V^{n+1} V^{*n+1}$$

implies (iv).

For any $g \in \mathcal{G}$, $\|V^{*n}g\|^2 = \langle V^n V^{*n}g, g\rangle \to \langle Pg, g\rangle = \|Pg\|^2$. Hence (v) holds.

For any $n = 0,1,2,\ldots$, $VV^n V^{*n} = V^{n+1} V^{*n+1} V$. Letting $n \to \infty$, we get $VP = PV$, so (vi) holds. Similarly, $VPV^* = P$, and (vii) follows.

Let $S = V|Q\mathcal{G}$. Then $S^{*n} = V^{*n}|Q\mathcal{G} \to 0$ strongly by (v). Thus S is a shift operator and (viii) holds.

The first relation in (ix) follows from (iv). Arguing as in the proof of (i), we see that

$$V^j P_0 V^{*j} = V^j V^{*j} - V^{j+1} V^{*j+1}$$

is the projection of $\mathcal{G}$ on $V^j\mathcal{G} \ominus V^{j+1}\mathcal{G} = V^j P_0 \mathcal{G}$. Thus $P\mathcal{G}$, $P_0\mathcal{G}$, $VP_0\mathcal{G}$, $V^2 P_0 \mathcal{G},\ldots$ are orthogonal subspaces of $\mathcal{G}$ with associated projections P, P_0, $VP_0 V^*$, $V^2 P_0 V^{*2},\ldots$. Hence the second relation in (ix) follows, and this completes the proof. ∎

COROLLARY A. *An isometry $V \in \mathcal{B}(\mathcal{G})$ is a shift operator if and only if $\bigcap_0^\infty V^j\mathcal{G} = \{0\}$.*

Specializing the Wold decomposition to the case of a shift operator, we obtain:

COROLLARY B. *If $S \in \mathcal{B}(\mathcal{H})$ is a shift operator and $\mathcal{K} = \ker S^*$, then $\mathcal{H} = \sum_0^\infty \oplus S^j\mathcal{K}$. Each $f \in \mathcal{H}$ has a unique representation*

$$f = \sum_0^\infty S^j k_j, \tag{1-4}$$

where $k_j \in \mathcal{K}, j \geq 0$. In this case, $\|f\|^2 = \sum_0^\infty \|k_j\|^2$ and

$$k_j = P_0 S^{*j} f, \qquad j \geq 0, \tag{1-5}$$

where $P_0 = I - SS^*$ is the projection of $\mathcal{H}$ on $\mathcal{K}$.

When S is multiplication by z on $H^2(D)$, $\mathcal{K} = \ker S^*$ is the set of constant functions in $H^2(D)$. If we identify $\mathcal{K}$ with $\mathbf{C}$ in the obvious way, then the expansion (1-4) of any function in $H^2(D)$ takes the form $f(z) = \sum_0^\infty a_j z^j$ and coincides with the Taylor series representation.

COROLLARY C.   *Let $S \in \mathcal{B}(\mathcal{H})$ be a shift operator. A subspace M of $\mathcal{H}$ reduces S if and only if*

$$M = \sum_0^\infty \oplus S^j M_0, \tag{1-6}$$

where M_0 is a subspace of $\mathcal{K}$ = ker S^*.

Proof.   If M reduces S, then $S_M = S|M$ is a shift operator on M and $S_M^* = S^*|M$. Let $M_0 = \ker S_M^*$. Then $M_0 \subseteq \ker S^* = \mathcal{K}$, and (1-6) follows from Corollary B. The converse is clear.   ∎

1.4   Multiplicity

Let $\mathcal{H}$ be a Hilbert space. A subspace $\mathcal{E}$ of $\mathcal{H}$ is called *cyclic* for an operator $A \in \mathcal{B}(\mathcal{H})$ if $\bigvee_0^\infty A^j \mathcal{E} = \mathcal{H}$.

THEOREM A.   *If $S \in \mathcal{B}(\mathcal{H})$ is a shift operator, then $\mathcal{K} = \ker S^*$ is cyclic for S, and* dim $\mathcal{K} \le$ dim $\mathcal{E}$ *for every cyclic subspace $\mathcal{E}$ for S.*

LEMMA.   *Let $\mathcal{H}_1$ and $\mathcal{H}_2$ be two Hilbert spaces. If there exists a one-to-one operator $A \in \mathcal{B}(\mathcal{H}_1, \mathcal{H}_2)$, then* dim $\mathcal{H}_1 \le$ dim $\mathcal{H}_2$.

Proof of lemma.   Let $A = WB$ be the polar decomposition of A, so that B is a nonnegative operator in $\mathcal{B}(\mathcal{H}_1)$ and W is a partial isometry on $\mathcal{H}_1$ to $\mathcal{H}_2$ with initial space $\overline{B\mathcal{H}_1}$ and final space $\overline{A\mathcal{H}_1}$. Since ker $B = $ ker $A = \{0\}$, we have $(B\mathcal{H}_1)^\perp = \ker B^* = \ker B = \{0\}$ and $\overline{B\mathcal{H}_1} = \mathcal{H}_1$. Thus W is an isometry on $\mathcal{H}_1$ to $\mathcal{H}_2$. If $\{e_j\}_{j \in J}$ is an orthonormal basis for $\mathcal{H}_1$, then $\{We_j\}_{j \in J}$ is an orthonormal set in $\mathcal{H}_2$. Hence dim $\mathcal{H}_1 \le$ dim $\mathcal{H}_2$.   ∎

Proof of Theorem A.   Corollary B of Section 1.3 implies that $\mathcal{K}$ is cyclic for S. Let $\mathcal{E}$ be any cyclic subspace for S. If P_0 is the projection of $\mathcal{H}$ on $\mathcal{K}$, then $T = P_0|\mathcal{E}$ is in $\mathcal{B}(\mathcal{E}, \mathcal{K})$. Claim: $\overline{T\mathcal{E}} = \mathcal{K}$. To see this, consider any $k \in \mathcal{K} \ominus T\mathcal{E}$. For all $e \in \mathcal{E}$,

$$\langle e, k \rangle = \langle e, P_0 k \rangle = \langle P_0 e, k \rangle = 0$$

because $k \perp T\mathcal{E} = P_0\mathcal{E}$. Since $\mathcal{K} = \mathcal{H} \ominus S\mathcal{H}$, we also have $S^j e \perp k, j = 1, 2, 3, \ldots$. Thus $k \perp S^j \mathcal{E}$ for all $j = 0, 1, 2, \ldots$, and since $\mathcal{E}$ is cyclic, $k = 0$. Therefore $\overline{T\mathcal{E}} = \mathcal{K}$.

Now $T^* \in \mathcal{B}(\mathcal{K}, \mathcal{E})$ and ker $T^* = \mathcal{K} \ominus T\mathcal{E} = \{0\}$. Hence T^* is one-to-one, and by the lemma, dim $\mathcal{K} \le$ dim $\mathcal{E}$.   ∎

We define the *multiplicity* of a shift operator $S \in \mathcal{B}(\mathcal{H})$ to be the minimum dimension of a cyclic subspace for S. By Theorem A, the multiplicity of S is dim $\mathcal{K}$, where $\mathcal{K} = \ker S^*$. For any Hilbert space $\mathcal{C}$, the multiplicity of $S: (c_0, c_1, c_2, \ldots) \to (0, c_0, c_1, \ldots)$ on $l_\mathcal{C}^2$ is equal to the dimension of $\mathcal{C}$.

DEFINITION.   *An operator $A \in \mathcal{B}(\mathcal{H})$ is unitarily equivalent to an operator $B \in \mathcal{B}(\mathcal{K})$ if there is a Hilbert space isomorphism U mapping $\mathcal{H}$ onto $\mathcal{K}$ such that $A = U^{-1} BU$.*

Typically, in operator theory unitarily equivalent operators are viewed as being indistinguishable.

THEOREM B.   *Two shift operators are unitarily equivalent if and only if they have the same multiplicity.*

Proof.   Let $S_j \in \mathcal{B}(\mathcal{H}_j), j = 1,2$, be shift operators. If S_1 and S_2 have the same multiplicity, then the subspaces $\mathcal{K}_j = \ker S_j^*, j = 1,2$, have the same dimension. Hence there is an isometry W_0 which maps $\mathcal{K}_1$ onto $\mathcal{K}_2$. For any $f \in \mathcal{H}_1$, define

$$Wf = \sum_0^\infty S_2^j W_0 k_j \quad \text{if} \quad f = \sum_0^\infty S_1^j k_j$$

as in (1-4). Then W is an isomorphism from $\mathcal{H}_1$ to $\mathcal{H}_2$ such that $S_2 W = WS_1$. Thus S_1 and S_2 are unitarily equivalent.

In the other direction, if S_1 and S_2 are unitarily equivalent, then the kernels of S_1^* and S_2^* are isomorphic. Hence S_1 and S_2 have the same multiplicity.   ∎

1.5   Universal Model

Shift operators have the following remarkable property:

Up to unitary equivalence and multiplicative constants, the class of operators $T = S^*|M$, where S is a shift operator and M is an invariant subspace for S^*, includes every bounded linear operator on a Hilbert space.

More precisely:

THEOREM.   *Let T be a bounded linear operator on a Hilbert space $\mathcal{H}$ such that $\|T\| \leq 1$ and $\|T^n f\| \rightarrow 0$ for each $f \in \mathcal{H}$. Let S be a shift operator on a Hilbert space $\mathcal{G}$ of multiplicity $\geq \dim((I - T^*T)\mathcal{H})^-$. Then there exists an invariant subspace M of S^* such that T is unitarily equivalent to $S^*|M$.*

If $T \in \mathcal{B}(\mathcal{H})$ and T does not satisfy the hypotheses of the theorem, then cT will satisfy the hypotheses for any scalar $c \neq 0$ such that $\|cT\| < 1$. In this case, it is necessary to choose a shift operator S whose multiplicity is $\dim \mathcal{H}$.

Proof.   Let $\mathcal{K} = \ker S^*$. Our assumptions imply that

$$\dim((I - T^*T)^{1/2}\mathcal{H})^- = \dim((I - T^*T)\mathcal{H})^- \leq \dim \mathcal{K}.$$

Hence there is an isometry J that maps $((I - T^*T)^{1/2}\mathcal{H})^-$ into $\mathcal{K}$. Define $W: \mathcal{H} \rightarrow \mathcal{G}$ by

$$Wf = \sum_0^\infty S^j J(I - T^*T)^{1/2} T^j f, \qquad f \in \mathcal{H}.$$

By Corollary B of Section 1.3, for any $f \in \mathcal{H}$,

$$\|Wf\|^2 = \sum_0^\infty \|J(I - T^*T)^{1/2}T^j f\|_{\mathcal{G}}^2$$

$$= \sum_0^\infty \|(I - T^*T)^{1/2}T^j f\|_{\mathcal{H}}^2$$

$$= \lim_{n \to \infty} \sum_0^n \langle T^{*j}(I - T^*T)T^j f, f \rangle_{\mathcal{H}}$$

$$= \lim_{n \to \infty} (\|f\|_{\mathcal{H}}^2 - \|T^{n+1}f\|_{\mathcal{H}}^2)$$

$$= \|f\|_{\mathcal{H}}^2.$$

Hence W is an isometry on $\mathcal{H}$ to $\mathcal{G}$. Let $M = W\mathcal{H}$. Then W is a Hilbert space isomorphism of $\mathcal{H}$ onto M. For each $f \in \mathcal{H}$,

$$S^*Wf = \sum_0^\infty S^j J(I - T^*T)^{1/2}T^j(Tf) = WTf.$$

It follows that M is invariant under S^*, and T is unitarily equivalent to $S^*|M$. ∎

1.6   Analytic Operators

Let $S \in \mathcal{B}(\mathcal{H})$ be a shift operator. An operator $A \in \mathcal{B}(\mathcal{H})$ is

  (i) *S-analytic* if $AS = SA$,
 (ii) *S-inner* if A is analytic and partially isometric, and
(iii) *S-outer* if A is analytic and $\overline{A\mathcal{H}}$ reduces S.

An analytic operator $A \in \mathcal{B}(\mathcal{H})$ is said to be *S-constant* if A^* is also analytic.

The terminology *analytic,* *inner,* and *outer* is also used when there is no possibility of confusion. To justify the terminology, consider the example where S is multiplication by z on $H^2(D)$.

  (a) A bounded linear operator A on $H^2(D)$ is *S*-analytic if and only if $Af = af$, $f \in H^2(D)$, for some $a \in H^\infty(D)$. In this case, $\|A\| = \|a\|_\infty$.

In (b)–(d) below assume that $Af = af$, $f \in H^2(D)$, for a fixed $a \in H^\infty(D)$, $a \not\equiv 0$. Then:

  (b) A is *S*-inner if and only if a is an inner function.
  (c) A is *S*-outer if and only if a is an outer function.
  (d) A is *S*-constant if and only if $a \equiv$ const. on D.

Assertions (a) and (d) are special cases of 1.15, Theorem B. We leave (b) and (c) to the reader (see p. 95).

1.7   Inner Operators

Let $S \in \mathscr{B}(\mathscr{H})$ be a shift operator, and let $\mathscr{K} = \ker S^*$.

THEOREM A.   *The initial space of any inner operator $B \in \mathscr{B}(\mathscr{H})$ reduces S.*

Proof.   The initial space of B is given by $M = \{f \in \mathscr{H} : \|Bf\| = \|f\|\}$. If $f \in \mathscr{H}$ and $\|Bf\| = \|f\|$, then

$$\|BSf\| = \|SBf\| = \|Bf\| = \|f\| = \|Sf\|.$$

Hence M is invariant under S. Since $M^{\perp} = \ker B$ and $BS = SB$, $M^{\perp}$ is also invariant under S. Thus M reduces S. ∎

We next describe all of the S-constant inner operators on $\mathscr{H}$. To construct an example, choose a partial isometry $B_0 \in \mathscr{B}(\mathscr{K})$. By Section 1.3, Corollary B, each $f \in \mathscr{H}$ has the form $f = \sum_0^\infty S^j k_j$, where $\{k_j\}_0^\infty \subseteq \mathscr{K}$. Define an operator $B \in \mathscr{B}(\mathscr{H})$ by setting

$$Bf = \sum_0^\infty S^j B_0 k_j \tag{1-7}$$

in this situation. It is easy to see that B is inner. Moreover,

$$B^*f = \sum_0^\infty S^j B_0^* k_j$$

if $f = \sum_0^\infty S^j k_j$ as above. Hence B^* is also inner, and B is S-constant. This example is general.

THEOREM B.   *Every S-constant inner operator $B \in \mathscr{B}(\mathscr{H})$ has the form just described for some partial isometry $B_0 \in \mathscr{B}(\mathscr{K})$.*

Proof.   First note that $\mathscr{K}$ reduces B. For since $BS^* = S^*B$, $B\mathscr{K} \subseteq \mathscr{K}$, and since $BS = SB$, $\mathscr{K}^{\perp} = S\mathscr{H}$ is also invariant under B. Therefore the projection P_0 of $\mathscr{H}$ on $\mathscr{K}$ commutes with B, and hence P_0 also commutes with B^*.

Define $B_0 = B|\mathscr{K}$. Clearly (1-7) holds whenever $f = \sum_0^\infty S^j k_j$, $\{k_j\}_0^\infty \subseteq \mathscr{K}$. It remains only to show that B_0 is a partial isometry. By Section 1.2, B^*B is a projection operator. Since B^*B and P_0 commute, $(BP_0)^*(BP_0) = B^*B \cdot P_0$ is a projection operator (Halmos [1951], p. 47). Again by Section 1.2, BP_0 is a partial isometry on $\mathscr{H}$, and from this we readily deduce that B_0 is a partial isometry on $\mathscr{K}$ to $\mathscr{K}$. ∎

THEOREM C.   *The final space of an inner operator $B \in \mathscr{B}(\mathscr{H})$ reduces S if and only if B is S-constant.*

Proof.   If B is S-constant, then B^* is also inner. The final space for B is the initial space for B^*. Hence the sufficiency part follows from Theorem A.

Conversely, suppose that the final space N of B reduces S. By Section 1.2, BB^* is the projection of $\mathscr{H}$ on N. Since N reduces S, $S(BB^*) = (BB^*)S$. Therefore $B(SB^* - B^*S) = 0$ and $(SB^* - B^*S)\mathscr{H} \subseteq \ker B$. Claim: $(SB^* - B^*S)\mathscr{H} \perp$

ker B. For if $u \in \ker B$, then $S^*u \in \ker B$ by Theorem A. Hence for any $f \in \mathscr{H}$,

$$\langle (SB^* - B^*S)f,u \rangle = \langle f,BS^*u \rangle - \langle Sf,Bu \rangle = 0.$$

The claim follows. Then

$$(SB^* - B^*S)\mathscr{H} \subseteq \ker B \cap (\ker B)^{\perp} = \{0\},$$

so $SB^* = B^*S$. Thus B^* is analytic, and so B is S-constant.   ∎

1.8   Two Factorization Problems

The following questions are basic. Let $S \in \mathscr{B}(\mathscr{H})$ be a shift operator.

PROBLEM 1. (*Abstract Beurling Problem*).   *Characterize all products* AA^*, *where A is an S-analytic operator on $\mathscr{H}$.*

PROBLEM 2. (*Abstract Szegö Problem*).   *Characterize all products* A^*A, *where A is an S-analytic operator on $\mathscr{H}$.*

We will solve Problem 1 in this chapter and apply the solution to prove the Beurling-Lax Theorem 1.12. This leads to an inner-outer factorization theory for analytic operators (Chapter 3), and hence to an inner-outer factorization theory for operator valued holomorphic functions (Chapter 5). Problem 2 is treated in Chapter 3. It is also intimately connected with the inner-outer factorization theory. More precisely, the applications of Problem 2 yield a factorization theory for nonnegative operator valued functions on a circle or line (Chapter 6).

1.9   Theorem. Solution of Problem 1

Let $S \in \mathscr{B}(\mathscr{H})$ be a shift operator, and let $\mathscr{K} = \ker S^*$. If $T \in \mathscr{B}(\mathscr{H})$, then the following are equivalent:

   (i) $T = AA^*$ *for some S-analytic operator* $A \in \mathscr{B}(\mathscr{H})$;
   (ii) $T - STS^* = J^*J$ *for some operator* $J \in \mathscr{B}(\mathscr{H},\mathscr{K})$;
   (iii) $T - STS^* \geq 0$ *and the rank of $T - STS^*$ does not exceed the multiplicity of S.*

The rank of an operator is the dimension of the closure of its range.

Proof.   (i) ⇔ (ii) If $T = AA^*$ where A is S-analytic, then

$$T - STS^* = A(I - SS^*)A^* = AP_0A^* = J^*J,$$

where $P_0 = I - SS^*$ is the projection of $\mathscr{H}$ on $\mathscr{K}$ and $J = P_0A^* \in \mathscr{B}(\mathscr{H},\mathscr{K})$.

Conversely, let $T - STS^* = J^*J$, where $J \in \mathscr{B}(\mathscr{H},\mathscr{K})$. Repeated application of the equation $T - STS^* = J^*J$ yields

$$T - S^{n+1}TS^{*n+1} = \sum_{0}^{n} S^j J^* J S^{*j},$$

$n = 0,1,2,\ldots$. Viewing J as an operator on $\mathscr{H}$ to $\mathscr{H}$, we obtain

$$\langle Tf,g\rangle - \langle S^{n+1}TS^{*n+1}f,g\rangle = \sum_0^n \langle JS^{*j}f, JS^{*j}g\rangle$$

$$= \left\langle \sum_0^n S^j JS^{*j}f, \sum_0^n S^j JS^{*j}g\right\rangle$$

for all $f, g \in \mathscr{H}$ and $n = 0,1,2,\ldots$. Define $A \in \mathscr{B}(\mathscr{H})$ so that $A^* = \sum_0^\infty S^j JS^{*j}$. It is easy to see that the series for A^* converges strongly and A is S-analytic. Letting $n \to \infty$ in the preceding identity, we obtain $\langle Tf,g\rangle = \langle A^*f, A^*g\rangle$ for all $f, g \in \mathscr{H}$, so $T = AA^*$.

(ii) $\Leftrightarrow$ (iii) Let $T - STS^* = J^*J$, where $J \in \mathscr{B}(\mathscr{H},\mathscr{K})$. Clearly $T - STS^* \geq 0$. Let $J = WB$ be the polar decomposition of J. Thus $B = (J^*J)^{1/2}$ and W is a partial isometry on $\mathscr{H}$ to $\mathscr{K}$ with initial space $\overline{B\mathscr{H}}$. The range of $T - STS^*$ is contained in $B\mathscr{H}$ since $T - STS^* = J^*J = B^2$. Since W maps $\overline{B\mathscr{H}}$ isometrically into $\mathscr{K}$, the rank of $T - STS^*$ does not exceed $\dim \mathscr{K}$, which is the multiplicity of S. Hence (ii) implies (iii).

Conversely, let (iii) hold, and set $B = (T - STS^*)^{1/2}$. Since the range of B and the range of $B^2 = T - STS^*$ have the same closure, $\dim \overline{B\mathscr{H}} \leq \dim \mathscr{K}$. Therefore there is an isometry W in $\mathscr{B}(B\overline{\mathscr{H}},\mathscr{K})$. Then $J = WB \in \mathscr{B}(\mathscr{H},\mathscr{K})$ and $T - STS^* = J^*J$; that is, (ii) holds. ∎

1.10 A Uniqueness Theorem

Let $S \in \mathscr{B}(\mathscr{H})$ be a shift operator, let $\mathscr{K} = \ker S^*$, and let $P_0 = I - SS^*$ be the projection of $\mathscr{H}$ on $\mathscr{K}$.

By the *support* of an S-analytic operator $A \in \mathscr{B}(\mathscr{H})$ we mean the smallest reducing subspace $M(A)$ for S containing $\overline{A^*\mathscr{H}} = (\ker A)^\perp$. Equivalently, $M(A)$ is the smallest reducing subspace N for S such that $A|N^\perp = 0$. Thus $M(A)$ reduces S, $A|M(A)^\perp = A^*|M(A)^\perp = 0$, and no proper subspace of $M(A)$ has these properties. We show that

$$M(A) = \sum_0^\infty \oplus\, S^j M_0(A), \qquad (1\text{-}8)$$

where $M_0(A) = \overline{P_0 A^*\mathscr{H}}$. Indeed, $M(A)$ contains $(I - SS^*)A^*\mathscr{H} = P_0 A^*\mathscr{H}$, and so

$$M(A) \supseteq \sum_0^\infty \oplus\, S^j M_0(A).$$

The direct sum on the right reduces S and contains $A^*\mathscr{H}$. For if $f \in \mathscr{H}$ and $A^*f = \sum_0^\infty S^j k_j$ as in (1-4), then for all $j \geq 0$, by (1-5),

$$k_j = P_0 S^{*j} A^* f = P_0 A^* S^{*j} f \in M_0(A).$$

Thus the direct sum contains $M(A)$ and (1-8) holds.

THEOREM.   *Let $A,C \in \mathscr{B}(\mathscr{H})$ be S-analytic. Then*

$$AA^* = CC^*$$

if and only if $C = AB$, where B is an S-constant inner operator with initial space $M(C)$ and final space $M(A)$. In this case B is unique and $A = CB^*$.

Proof.   Suppose that $AA^* = CC^*$, and denote this operator T. Then

$$CP_0C^* = T - STS^* = AP_0A^*,$$

and for any $f \in \mathscr{H}$,

$$\|P_0C^*f\|^2 = \langle CP_0C^*f, f \rangle = \langle AP_0A^*f, f \rangle = \|P_0A^*f\|^2.$$

It follows that $P_0A^* = B_0P_0C^*$ for a unique partial isometry $B_0 \in \mathscr{B}(\mathscr{H})$ with initial space $M_0(C) = \overline{P_0C^*\mathscr{H}}$ and final space $M_0(A) = \overline{P_0A^*\mathscr{H}}$. As in (1-7) we extend B_0 to an S-constant inner operator B on $\mathscr{H}$ with initial space $M(C)$ and final space $M(A)$. We show that $C = AB$. Since $CP_0C^* = AP_0A^* = AB_0P_0C^*$, $C|M_0(C) = AB_0|M_0(C)$. Both C and AB_0 are zero on $\mathscr{K} \ominus M_0(C)$, so $C|\mathscr{K} = AB_0|\mathscr{K}$. If $f \in \mathscr{H}$ and $f = \sum_0^\infty S^jk_j$ as in (1-4), we then have

$$Cf = \sum_0^\infty S^jCk_j = \sum_0^\infty S^jAB_0k_j = ABf,$$

and so $C = AB$.

The operator B is unique. If $C = AB$ as in the theorem, then $B^*A^* = C^*$ and B^* is determined on $A^*\mathscr{H}$. Hence B^* is determined on $M(A)$. Uniqueness follows.

Now assume that $C = AB$, where B is an S-constant inner operator with initial space $M(C)$ and final space $M(A)$. By Section 1.2, BB^* is the projection of $\mathscr{H}$ on $M(A)$, and so BB^* coincides with the identity operator on $A^*\mathscr{H}$. Therefore

$$CC^* = ABB^*A^* = AA^*.$$

Similarly,

$$BC^* = BB^*A^* = A^*.$$

Hence $A = CB^*$, and the proof is complete.   ■

1.11   Counting Lemma

For any projection P on a separable Hilbert space $\mathscr{H}$ and any orthonormal basis $\{e_j\}_{j \in J}$ for $\mathscr{H}$,

$$\dim P\mathscr{H} = \sum_{j \in J} \|Pe_j\|^2.$$

Proof.   Let $\{f_k\}_{k \in K}$ be an orthonormal basis for $P\mathscr{H}$. Then

$$\sum_{j \in J} \|Pe_j\|^2 = \sum_{j \in J}\sum_{k \in K} |\langle Pe_j, f_k \rangle|^2$$

$$= \sum_{k \in K}\sum_{j \in J} |\langle e_j, f_k \rangle|^2$$

$$= \sum_{k \in K} \|f_k\|^2$$

$$= \dim P\mathscr{H},$$

whether the sums are finite or not. The separability assumption is used only for the last equality when the sums are infinite. ∎

1.12 Beurling-Lax Theorem

Let S be a shift operator on a Hilbert space $\mathcal{H}$. A subspace M of $\mathcal{H}$ is invariant under S if and only if $M = A\mathcal{H}$ for some S-inner operator A on $\mathcal{H}$.

This representation of an invariant subspace is essentially unique. Suppose that an invariant subspace M of S is represented as $M = A\mathcal{H}$ and $M = C\mathcal{H}$ for two S-inner operators A and C. Then $AA^* = CC^*$, so by Section 1.10,

$$C = AB \qquad \text{and} \qquad A = CB^*,$$

where B is an S-constant inner operator whose initial space is the support of C and whose final space is the support of A. Conversely,

$$A\mathcal{H} = C\mathcal{H}$$

whenever A and C are S-inner operators related in this way.

Proof. If $M = A\mathcal{H}$, where A is S-inner, then $SM = SA\mathcal{H} = AS\mathcal{H} \subseteq A\mathcal{H} = M$.

Conversely, assume that M is invariant under S. Let P be the projection of $\mathcal{H}$ on M. Then SPS^* is the projection of $\mathcal{H}$ on SM and $Q = P - SPS^*$ is the projection of $\mathcal{H}$ on $M \ominus SM$.

We show that the dimension of $Q\mathcal{H}$ does not exceed the multiplicity of S, that is, $\dim Q\mathcal{H} \leq \dim \mathcal{K}$, where $\mathcal{K} = \ker S^*$. If $\mathcal{K}$ is infinite dimensional of any cardinality, then $\dim Q\mathcal{H} \leq \dim \mathcal{H} = \dim \mathcal{K}$ (the equality is by Corollary B of Section 1.3). Let $\mathcal{K}$ be finite dimensional with orthonormal basis $\{e_k\}_{k \in K}$. Then $\{S^j e_k : k \in K, j = 0,1,2,\ldots\}$ is an orthonormal basis for $\mathcal{H}$. In this case $\mathcal{H}$ is separable and by Section 1.11,

$$\dim Q\mathcal{H} = \sum_{k \in K} \sum_{j=0}^{\infty} \langle QS^j e_k, S^j e_k \rangle$$

$$= \lim_{n \to \infty} \sum_{k \in K} \sum_{j=0}^{n} \langle (P - SPS^*)S^j e_k, S^j e_k \rangle$$

$$= \lim_{n \to \infty} \sum_{k \in K} \langle PS^n e_k, S^n e_k \rangle$$

$$\leq \sum_{k \in K} \|e_k\|^2$$

$$= \dim \mathcal{K},$$

as required.

It follows from what we have shown that P satisfies condition (iii) of Section 1.9. Therefore by 1.9, $P = AA^*$ for some S-analytic operator A. Since P is a projection, A is partially isometric and hence S-inner. By construction, $M = P\mathcal{H} = A\mathcal{H}$, and this completes the proof. ∎

The commutant of an operator $T \in \mathscr{B}(\mathscr{H})$ is the set $C(T)$ of all $X \in \mathscr{B}(\mathscr{H})$ such that $XT = TX$. More generally, if $T_1 \in \mathscr{B}(\mathscr{H}_1)$ and $T_2 \in \mathscr{B}(\mathscr{H}_2)$, let $C(T_1,T_2)$ be the set of all $X \in \mathscr{B}(\mathscr{H}_1,\mathscr{H}_2)$ such that $XT_1 = T_2X$. The lifting theorem 1.13 characterizes $C(T_1,T_2)$ when T_1 and T_2 are represented as in the universal model 1.5.

1.13   Lifting Theorem

For $j = 1,2$, let $S_j \in \mathscr{B}(\mathscr{G}_j)$ be a shift operator, let $\mathscr{H}_j$ be an invariant subspace for S_j^*, and let $T_j = S_j^* | \mathscr{H}_j$. Let $X \in \mathscr{B}(\mathscr{H}_1,\mathscr{H}_2)$ satisfy

$$XT_1 = T_2X. \tag{1-9}$$

Then there exists an operator $Y \in \mathscr{B}(\mathscr{G}_1,\mathscr{G}_2)$ such that (i) $Y\mathscr{H}_1 \subseteq \mathscr{H}_2$ and $X = Y | \mathscr{H}_1$, (ii) $YS_1^* = S_2^* Y$, and (iii) $\|Y\| = \|X\|$.

We shall deduce 1.13 from a more general result, 1.14. Before studying the proof of 1.14, we recommend that the reader first see how 1.13 follows from 1.14.

1.14   Theorem. Abstract Form of the Lifting Theorem

Let $S \in \mathscr{B}(\mathscr{H})$ be an isometry, and let $T, R \in \mathscr{B}(\mathscr{K})$ be operators such that $T \geq 0$ and $T \leq R^* TR$. Let M be an invariant subspace for S^*, and let P be the projection of $\mathscr{H}$ on M. Let M' be an invariant subspace for R^*. Let $X \in \mathscr{B}(\mathscr{K},\mathscr{H})$ satisfy

   (i) $XM' \subseteq M$ and $XM'^{\perp} = \{0\}$,
   (ii) $PSX = XR$,
   (iii) $X^*X \leq T$.

Then there exists an operator $Y \in \mathscr{B}(\mathscr{K},\mathscr{H})$ such that

   (i′) $X = PY$,
   (ii′) $SY = YR$,
   (iii′) $Y^*Y \leq T$.

   LEMMA.   *Let $A \in \mathscr{B}(\mathscr{H}_1,\mathscr{H}_3)$, $C \in \mathscr{B}(\mathscr{H}_1,\mathscr{H}_2)$, and $\beta > 0$ be given. The following are equivalent:*

   (i) *$A = BC$ for some $B \in \mathscr{B}(\mathscr{H}_2,\mathscr{H}_3)$ such that $\|B\| \leq \beta$;*
   (ii) *$A^*A \leq \beta^2 C^*C$.*

   Proof of lemma.   Assume (ii). For each $f \in \mathscr{H}_1$,

$$\|Af\|_3^2 = \langle A^*Af,f \rangle_1 \leq \beta^2 \langle C^*Cf,f \rangle_1 = \beta^2\|Cf\|_2^2.$$

Hence we may define $B_0 \colon \overline{C\mathscr{H}_1} \to \mathscr{H}_3$ by $B_0(Cf) = Af$, $f \in \mathscr{H}_1$. We have $\|B_0\| \leq \beta$. Extend B_0 to an operator $B \in \mathscr{B}(\mathscr{H}_2,\mathscr{H}_3)$ such that B is zero on $\mathscr{H}_2 \ominus C\mathscr{H}_1$. Then $A = BC$ and $\|B\| \leq \beta$; that is, (i) follows.
   The other direction is straightforward.   ■

Proof of Theorem 1.14.   Let $Q = I - P$. Then $SQ\mathscr{H} \subseteq Q\mathscr{H}$, and $S|Q\mathscr{H}$ is an isometry. Let P_0 be the projection of $\mathscr{H}$ on $\ker((S|Q\mathscr{H})^*) = Q\mathscr{H} \ominus SQ\mathscr{H}$. Thus $P_0 = Q - SQS^*$, and for $j,k = 0,1,2,\ldots,$

$$P_0 S^{*k} S^j P_0 = \begin{cases} 0 & \text{if } j \neq k, \\ P_0 & \text{if } j = k, \end{cases} \tag{1-10}$$

and

$$P_0 S^{*j} X = 0 \qquad \text{and} \qquad X^* S^j P_0 = 0. \tag{1-11}$$

We inductively construct sequences $\{B_n\}^{\infty}_{-1}$ and $\{Y_n\}^{\infty}_{-1} \subseteq \mathscr{B}(\mathscr{K},\mathscr{H})$ such that

$$Y_{-1} = X,$$

$$Y_n = X + \sum_0^n S^j B_j, \quad n = 0,1,2,\ldots.$$

We require for $n = -1,0,1,2,\ldots$ that

$$\alpha(n): \quad B_n \in \mathscr{B}(\mathscr{K}, P_0\mathscr{H}),$$

$$\beta(n): \quad Y_n^* Y_n - R^* Y_n^* Y_n R = B_n^* B_n,$$

$$\gamma(n): \quad Y_n^* Y_n \leq T \leq R^* TR,$$

and for $n = 0,1,2,\ldots$ that

$$\delta(n): \quad B_{n-1} = B_n R.$$

Let $B_{-1} = QSX$. Since $QS^* QSX = QS^*(I - P)SX = QX - QS^* XR = 0 - 0 = 0$,

$$P_0 B_{-1} = (Q - SQS^*)QSX = QSX = B_{-1}.$$

Thus $\alpha(-1)$ holds. Also,

$$X^*X - R^*X^*XR = X^*X - X^*S^*PSX = X^*S^*QSX = B^*_{-1}B_{-1},$$

so $\beta(-1)$ holds. The two inequalities in $\gamma(-1)$ hold by assumption.

Suppose that $B_{-1}, B_0, \ldots, B_n$ have been constructed for some $n \geq -1$. Then

$$B_n^* B_n = Y_n^* Y_n - R^* Y_n^* Y_n R \leq R^*(T - Y_n^* Y_n)R.$$

By the lemma there exists an operator $C_{n+1} \in \mathscr{B}(\mathscr{K}, P_0\mathscr{H})$ such that $\|C_{n+1}\| \leq 1$ and

$$B_n = C_{n+1}(T - Y_n^* Y_n)^{1/2} R.$$

Let

$$B_{n+1} = C_{n+1}(T - Y_n^* Y_n)^{1/2}.$$

Clearly $\alpha(n + 1)$ and $\delta(n + 1)$ hold. Hence $B_{n+1} = P_0 B_{n+1}$, and by (1-11),

$$B^*_{n+1} S^{*n+1} Y_n = B^*_{n+1} P_0 S^{*n+1}(X + \sum_0^n S^j B_j) = 0.$$

Therefore

$$
\begin{aligned}
Y^*_{n+1} & Y_{n+1} - R^* Y^*_{n+1} Y_{n+1} R \\
&= (Y^*_n + B^*_{n+1} S^{*n+1})(Y_n + S^{n+1} B_{n+1}) \\
&\quad - R^*(Y^*_n + B^*_{n+1} S^{*n+1})(Y_n + S^{n+1} B_{n+1})R \\
&= Y^*_n Y_n + B^*_{n+1} B_{n+1} - R^*(Y^*_n Y_n + B^*_{n+1} B_{n+1})R \\
&= (Y^*_n Y_n - R^* Y^*_n Y_n R) + (B^*_{n+1} B_{n+1} - R^* B^*_{n+1} B_{n+1} R) \\
&= B^*_n B_n + (B^*_{n+1} B_{n+1} - B^*_n B_n) \\
&= B^*_{n+1} B_{n+1},
\end{aligned}
$$

so $\beta(n + 1)$ holds. Similarly,

$$
\begin{aligned}
Y^*_{n+1} Y_{n+1} &= Y^*_n Y_n + (T - Y^*_n Y_n)^{1/2} C^*_{n+1} C_{n+1} (T - Y^*_n Y_n)^{1/2} \\
&\leq Y^*_n Y_n + (T - Y^*_n Y_n) \\
&= T,
\end{aligned}
$$

and $\gamma(n + 1)$ follows. This completes the inductive construction.

It follows from (1-10) that $\{Y_n\}_0^\infty$ converges strongly to an operator $Y \in \mathcal{B}(\mathcal{K}, \mathcal{H})$. Thus

$$
Y = X + \sum_0^\infty S^j B_j,
$$

where the series converges strongly. The assertions (i′) and (iii′) are immediate. For each $n = 1, 2, \ldots,$

$$
\begin{aligned}
Y_n R &= XR + \sum_0^n S^j B_j R \\
&= XR + B_{-1} + \sum_0^{n-1} S^{j+1} B_j \\
&= PSX + QSX + \sum_0^{n-1} S^{j+1} B_j \\
&= SX + \sum_0^{n-1} S^{j+1} B_j \\
&= SY_{n-1}.
\end{aligned}
$$

Thus (ii′) holds, and the result follows.   ∎

Proof of Theorem 1.13.   First change the notation in Theorem 1.14. In 1.14 write $\tilde{X}, \tilde{Y}$ in place of X, Y.

We apply 1.14 with $S = S_1$ on $\mathcal{H} = \mathcal{G}_1$, $R = S_2$ on $\mathcal{K} = \mathcal{G}_2$, $M = \mathcal{H}_1$, $M' = \mathcal{H}_2$, and $T = \|X\|^2 I_2$, where I_2 is the identity operator on $\mathcal{G}_2$. Choose $\tilde{X} \in \mathcal{B}(\mathcal{G}_2, \mathcal{G}_1)$ so that $\tilde{X}|\mathcal{H}_2 = X^*$ and $\tilde{X}|\mathcal{H}_2^\perp = 0$. The hypotheses of 1.14 are then satisfied. Let $\tilde{Y}$ be the operator produced by 1.14. Define $Y \in \mathcal{B}(\mathcal{G}_1, \mathcal{G}_2)$ so that $\tilde{Y} = Y^*$. It is easy to see that Y has the required properties.   ∎

1.15 Concrete Realization of a Shift Operator

Let $\mathscr{C}$ be a Hilbert space with inner product $\langle \cdot, \cdot \rangle_\mathscr{C}$ and norm $|\cdot|_\mathscr{C}$. The norm on $\mathscr{B}(\mathscr{C})$ is denoted $|\cdot|_{\mathscr{B}(\mathscr{C})}$.

DEFINITION. *By $H_\mathscr{C}^2(D)$ we mean the space of all $\mathscr{C}$-valued holomorphic functions $f(z) = \sum_0^\infty a_j z^j$ on D for which the quantity*

$$\frac{1}{2\pi} \int_0^{2\pi} |f(re^{i\theta})|_\mathscr{C}^2 \, d\theta = \sum_0^\infty |a_j|_\mathscr{C}^2 r^{2j}$$

remains bounded for $0 \le r < 1$.

For the rudiments of the theory of vector and operator valued holomorphic functions see Hille and Phillips [1957], Chapter III, §2. It is easy to see that $H_\mathscr{C}^2(D)$ is a Hilbert space with inner product given by

$$\langle f, g \rangle_2 = \lim_{r \uparrow 1} \frac{1}{2\pi} \int_0^{2\pi} \langle f(re^{i\theta}), g(re^{i\theta}) \rangle_\mathscr{C} \, d\theta = \sum_0^\infty \langle a_j, b_j \rangle_\mathscr{C}$$

for any $f(z) = \sum_0^\infty a_j z^j$ and $g(z) = \sum_0^\infty b_j z^j$ in the space. Thus $H_\mathscr{C}^2(D)$ is isomorphic with $l_\mathscr{C}^2$ via the correspondence between a function and its Taylor coefficients. As a consequence of this isomorphism, we obtain:

THEOREM A. *The operator multiplication by z on $H_\mathscr{C}^2(D)$, defined by S: $f(z) \to zf(z)$ for all $f(z)$ in $H_\mathscr{C}^2(D)$, is a shift operator of multiplicity $\dim \mathscr{C}$. The adjoint of S is $S^*: f(z) \to [f(z) - f(0)]/z$.*

COROLLARY. *Every shift operator on a Hilbert space is unitarily equivalent to multiplication by z on $H_\mathscr{C}^2(D)$ for some choice of $\mathscr{C}$.*

By $H_{\mathscr{B}(\mathscr{C})}^\infty(D)$ we mean the Banach algebra of bounded $\mathscr{B}(\mathscr{C})$-valued holomorphic functions A on D in the norm $\|A\|_\infty = \sup\{|A(z)|_{\mathscr{B}(\mathscr{C})}: z \in D\}$. Each $A \in H_{\mathscr{B}(\mathscr{C})}^\infty(D)$ induces an operator $T(A)$ on $H_\mathscr{C}^2(D)$ called *multiplication by A*, defined by

$$T(A): f \to Af, \qquad f \in H_\mathscr{C}^2(D).$$

THEOREM B. *Let S be multiplication by z on $H_\mathscr{C}^2(D)$. A bounded linear operator T on $H_\mathscr{C}^2(D)$ is S-analytic if and only if $T = T(A)$ for some $A \in H_{\mathscr{B}(\mathscr{C})}^\infty(D)$. In this case, $\|T\| = \|A\|_\infty$, and T is S-constant if and only if $A \equiv const.$*

LEMMA. *For all $f(z)$ in $H_\mathscr{C}^2(D)$, $c \in \mathscr{C}$, and $w \in D$,*

$$\langle f(z), c/(1 - \bar{w}z) \rangle_2 = \langle f(w), c \rangle_\mathscr{C}.$$

Proof of lemma. Compute the left and right sides of the identity in terms of the Taylor expansions of $f(z)$ and $c/(1 - \bar{w}z)$. ∎

Proof of Theorem B. If $T = T(A)$, where $A \in H_{\mathscr{B}(\mathscr{C})}^\infty(D)$, then it is clear that T is S-analytic and $\|T\| \le \|A\|_\infty$.

Conversely, assume that T is S-analytic. We may view any $c \in \mathscr{C}$ as a constant function in $H^2_{\mathscr{C}}(D)$. If $T: c \to f_c$, then for any $w \in D$ the mapping $A(w): c \to f_c(w)$ on $\mathscr{C}$ to $\mathscr{C}$ belongs to $\mathscr{B}(\mathscr{C})$ (use the lemma to prove boundedness). As a function of z, $A(z)$ is holomorphic on D. By construction, $T: c \to A(z)c$ for all $c \in \mathscr{C}$. Since $TS = ST$,

$$T: cz^j \to z^j A(z)c \tag{1-12}$$

for all $c \in \mathscr{C}$ and $j = 0,1,2,\ldots$. Every $f(z)$ in $H^2_{\mathscr{C}}(D)$ has a representation $f(z) = \sum_0^\infty a_j z^j$ that converges both pointwise on D and in the metric of $H^2_{\mathscr{C}}(D)$. Since T is continuous, by (1-12),

$$(Tf)(z) = \sum_0^\infty T\{a_j z^j\} = \sum_0^\infty z^j A(z)a_j = A(z)f(z). \tag{1-13}$$

Using the lemma and (1-13), we obtain

$$T^*: c/(1 - \bar{w}z) \to A(w)^* c/(1 - \bar{w}z) \tag{1-14}$$

for each $c \in \mathscr{C}$ and $w \in D$. Hence

$$|A(w)^* c|_{\mathscr{C}}/(1 - |w|^2) = \|A(w)^* c/(1 - \bar{w}z)\|_2^2$$
$$\leq \|T^*\|^2 \|c/(1 - \bar{w}z)\|_2^2$$
$$= \|T\|^2 |c|_{\mathscr{C}}^2/(1 - |w|^2).$$

Thus $A \in H^\infty_{\mathscr{B}(\mathscr{C})}(D)$ and $\|A\|_\infty \leq \|T\|$. By (1-13), $T = T(A)$. Hence $\|T\| \leq \|A\|_\infty$, and so $\|T\| = \|A\|_\infty$.

Suppose $T = T(A)$ is S-constant. Then $T^* = T(C)$ for some $C \in H^\infty_{\mathscr{B}(\mathscr{C})}(D)$ by what we just proved. By (1-14), $C(z) = A(w)^*$ for all $z, w \in D$. Hence $A(z) \equiv$ const. on D. Conversely, it is clear that if $A(z) \equiv$ const. on D, then $T = T(A)$ is S-constant. ∎

Hardy classes of vector valued functions are studied systematically in Chapter 4.

Examples and Addenda

1. The only reducing subspaces of a shift operator $S \in \mathscr{B}(\mathscr{H})$ of multiplicity 1 are $\{0\}$ and $\mathscr{H}$.

2. Operators $A \in \mathscr{B}(\mathscr{H})$ and $B \in \mathscr{B}(\mathscr{K})$ are called *similar* if $A = X^{-1}BX$ for some invertible operator $X \in \mathscr{B}(\mathscr{H},\mathscr{K})$. Two isometries that are similar are unitarily equivalent (Page [1970]).

3. (i) If M is an invariant subspace of a shift operator S, then $S|M$ is a shift operator of multiplicity not greater than the multiplicity of S.

(ii) Let $S \in \mathscr{B}(\mathscr{H})$ be a shift operator. If N is a subspace of $\mathscr{H}$ such that $S^j N \perp S^k N$ whenever $j \neq k$, $j,k = 0,1,2,\ldots$, then the dimension of N does not exceed the multiplicity of S.

4. Let $A = \begin{bmatrix} a & b \\ c & d \end{bmatrix}$, viewed as an operator on $\mathbf{C}^2$. Then

$$\|A\| = \tfrac{1}{2}N + \tfrac{1}{2}(N^2 - 4|D|^2)^{1/2},$$

where $N = |a|^2 + |b|^2 + |c|^2 + |d|^2$ and $D = ad - bc$. In particular, $\|A\| \leq 1$ if and only if $N \leq 1 + |D|^2$.

5. *Inequality of von Neumann and the invariant form of Schwarz's lemma.*
(i) Use the universal model in Section 1.5 to prove von Neumann's inequality: If $T \in \mathcal{B}(\mathcal{H})$ and $\|T\| \leq 1$, then $\|p(T)\| \leq 1$ for every polynomial $p(z)$ such that $|p(z)| \leq 1$ for $|z| \leq 1$.

(ii) Let $T = \begin{bmatrix} a & b \\ 0 & c \end{bmatrix}$ on $\mathcal{H} = \mathbf{C}^2$, where $a, c \in D$ and $|b|^2 = (1 - |a|^2)(1 - |c|^2)$.
Then $\|T\| = 1$ and for any polynomial $p(z)$,

$$p(T) = \begin{bmatrix} p(a) & b[p(a) - p(c)]/(a - c) \\ 0 & p(c) \end{bmatrix}.$$

Hence if $|p(z)| \leq 1$ for $|z| \leq 1$, then $\|p(T)\| \leq 1$ and so

$$\left| \frac{p(z) - p(w)}{z - w} \right|^2 \leq \frac{1 - |p(z)|^2}{1 - |z|^2} \frac{1 - |p(w)|^2}{1 - |w|^2}, \qquad z, w \in D.$$

(iii) Let $f(z)$ be holomorphic and satisfy $|f(z)| < 1$ on D. Use (ii) and an approximation argument to show that

$$\left| \frac{f(z) - f(w)}{z - w} \right|^2 \leq \frac{1 - |f(z)|^2}{1 - |z|^2} \frac{1 - |f(w)|^2}{1 - |w|^2}, \qquad z, w \in D.$$

Then use the identity $|1 - u\bar{v}|^2 = (1 - |u|^2)(1 - |v|^2) + |u - v|^2$ to deduce

$$\left| \frac{f(z) - f(w)}{1 - \overline{f(w)}\, f(z)} \right| \leq \left| \frac{z - w}{1 - \bar{w}z} \right|, \qquad z, w \in D.$$

Similar results are given by Williams [1967]. A connection between von Neumann's inequality and the Pick-Nevanlinna theorem is shown in Rovnyak [1982]. Foiaş [1957] has shown that von Neumann's inequality is false in general for every Banach space that is not a Hilbert space. See Burckel [1979] for the classical view of Schwarz's lemma.

6. *Laguerre shift.* The Laguerre polynomials of order 0 can be defined by either of the relations

$$e^{-t}L_n(t) = \frac{1}{n!}\left(\frac{d}{dt}\right)^n (t^n e^{-t}), \qquad n = 0, 1, 2, \ldots,$$

$$(1 - z)^{-1} e^{-tz/(1 - z)} = \sum_0^\infty L_n(t) z^n, \qquad |z| < 1.$$

For each $n = 0, 1, 2, \ldots,$

$$\int_0^x L_n(t)\, dt = L_n(x) - L_{n+1}(x), \tag{1-15}$$

$$\int_0^\infty e^{-st} e^{-\frac{1}{2}t} L_n(t)\, dt = (s - \tfrac{1}{2})^n / (s + \tfrac{1}{2})^{n+1}, \qquad \operatorname{Re} s > -\tfrac{1}{2}. \tag{1-16}$$

The functions $\{e^{-\frac{1}{2}t} L_n(t)\}_0^\infty$ form an orthonormal basis for $L^2(0,\infty)$ (Szegö [1975]).

THEOREM (von Neumann [1929b]).   (i) Let S be the shift operator on $L^2(0, \infty)$ such that

$$S: e^{-\frac{1}{2}t} L_n(t) \to e^{-\frac{1}{2}t} L_{n+1}(t), \qquad n \geq 0. \tag{1-17}$$

Then for each $f \in L^2(0, \infty)$,

$$S: f(x) \to f(x) - \int_0^x e^{-\frac{1}{2}(x-t)} f(t)\, dt. \tag{1-18}$$

(ii) Let T be the symmetric operator $i\, d/dx$ on $L^2(0, \infty)$, where the domain of T is taken as the set of (locally) absolutely continuous functions f on $(0, \infty)$ such that $f, f' \in L^2(0, \infty)$ and $f(x) \to 0$ as $x \downarrow 0$. Then $S = (T - \tfrac{1}{2}iI)(T + \tfrac{1}{2}iI)^{-1}$.

We call S the *Laguerre shift* on $L^2(0,\infty)$.

Proof. (i) By (1-15), (1-18) holds if $f(t) = e^{-\frac{1}{2}t} L_n(t)$ for some $n \geq 0$. The general case of (1-18) follows by linearity and approximation.

(ii) By the elementary theory of symmetric operators (Stone [1932], Chapter IX), S is the Cayley transform of the symmetric operator T_0 with graph

$$\mathscr{G}(T_0) = \{(f - Sf, \tfrac{1}{2}i(f + Sf)): f \in L^2(0, \infty)\}.$$

Thus $(p, q) \in \mathscr{G}(T_0)$ if and only if

$$p(x) = e^{-\frac{1}{2}x} \int_0^x e^{\frac{1}{2}t} f(t)\, dt,$$

$$q(x) = if(x) - \tfrac{1}{2}i e^{-\frac{1}{2}x} \int_0^x e^{\frac{1}{2}t} f(t)\, dt$$

for some $f \in L^2(0,\infty)$. A straightforward argument then shows that $\mathscr{G}(T_0)$ coincides with the graph of T, and (ii) follows.   ∎

7. *Shift operators and the Chebychev polynomials.*   The Chebychev polynomials $\{T_n(x)\}_0^\infty$ and $\{U_n(x)\}_0^\infty$ can be defined by the formal expansions

$$\frac{1 - xt}{1 - 2xt + t^2} = \sum_0^\infty T_n(x) t^n, \qquad \frac{1}{1 - 2xt + t^2} = \sum_0^\infty U_n(x) t^n.$$

For each $n \geq 0$, $T_n(\cos\theta) = \cos n\theta$, $U_n(\cos\theta) = \sin((n+1)\theta)/\sin\theta$. See Rivlin [1974] and Szegö [1975].

THEOREM A. *Let* $S \in \mathcal{B}(\mathcal{H})$ *be a shift operator. Write* $S = X + iY$, *where* $X = \operatorname{Re} S$, $Y = \operatorname{Im} S$, *and let* P_0 *be the projection on* $\mathcal{K} = \ker S^*$. *Then*

$$S^n P_0 = U_n(X)P_0, \qquad iYS^n P_0 = T_{n+1}(X)P_0 \tag{1-19}$$

for all $n \geq 0$, *and*

$$I = \sum_0^\infty U_n(X)P_0 U_n(X) \tag{1-20}$$

with convergence in the strong operator topology.

Proof. Obtain (1-19) by induction using the identities $U_{n+2}(x) = 2xU_{n+1}(x) - U_n(x)$, $n \geq 0$, and $T_{n+1}(x) = \frac{1}{2}[U_{n+1}(x) - U_{n-1}(x)]$, $n \geq 1$. Then (1-20) follows by the Wold decomposition. ∎

THEOREM B. *There is a unique shift operator* S_0 *on* $L^2(-1, 1)$ *such that*

$$S_0: (1 - x^2)^{1/4}U_n(x) \to (1 - x^2)^{1/4}U_{n+1}(x), \qquad n \geq 0.$$

For each $f \in L^2(-1, 1)$,

$$S_0: f(x) \to xf(x) - PV\frac{1}{\pi} \int_{-1}^{1} \frac{(1 - x^2)^{1/4}(1 - t^2)^{1/4}}{t - x} f(t)\,dt. \tag{1-21}$$

The real part of S_0, $X_0 = \operatorname{Re} S_0$, *is multiplication by* x *on* $L^2(-1, 1)$, *that is,* $X_0: f(x) \to xf(x)$.

Proof. The existence of S_0 follows from the fact that the functions $\{(2/\pi)^{1/2}(1 - x^2)^{1/4}U_n(x)\}_0^\infty$ form an orthonormal basis for $L^2(-1, 1)$. First check (1-21) on basis elements using the identity

$$PV\frac{1}{\pi} \int_{-1}^{1} \frac{(1 - t^2)^{1/2}U_n(t)}{t - x}\,dt = -T_{n+1}(x), \qquad |x| < 1$$

(Tricomi [1957], pp. 180–181), and then obtain the general case by linearity and approximation. ∎

THEOREM C. *The general form of a shift operator* S *on* $L^2(-1, 1)$ *whose real part* $X = \operatorname{Re} S$ *coincides with the real part* $X_0 = \operatorname{Re} S_0$ *of the operator (1-21) is*

$$S: f(x) \to xf(x) - PV\frac{1}{\pi} \int_{-1}^{1} \frac{(1 - x^2)^{1/4}(1 - t^2)^{1/4}}{t - x} \bar{C}(x)C(t)f(t)\,dt,$$

where $C(x)$ *is a measurable function such that* $|C(x)| = 1$ *a.e. on* $(-1, 1)$.

For details and a generalization to shift operators of higher multiplicity, see Rosenblum and Rovnyak [1978]. See also Suzuki [1976].

8. *The functional equation $g(x) - g(2x) = f(x)$.* We follow Rochberg [1968]. Let $\mathscr{H}_0$ be the Hilbert space of measurable complex valued functions $f(x)$ on $(-\infty, \infty)$ such that $f(x + 1) = f(x)$ a.e.,

$$\|f\|^2 = \int_0^1 |f(x)|^2 \, dx < \infty, \text{ and } \int_0^1 f(x) \, dx = 0.$$

(i) The operator $S_0: f(x) \to f(2x)$ on $\mathscr{H}_0$ is a shift operator with adjoint $S_0^*: f(x) \to \frac{1}{2} f(\frac{1}{2} x) + \frac{1}{2} f(\frac{1}{2} x + \frac{1}{2})$. If $\mathscr{K}_0 = \ker S_0^*$, then for each $n = 0, 1, 2, \ldots$,

$$S^n \mathscr{K}_0 = \bigvee \{\exp(2\pi i (2j + 1) 2^n x): j = 0, \pm 1, \pm 2, \ldots\}.$$

(ii) Let $S \in \mathscr{B}(\mathscr{H})$ be any shift operator, and let f be a vector in $\mathscr{H}$ for which the coefficients in the expansion $f = \sum_0^\infty S^j k_j$ of (1-4) satisfy

$$\|k_j\| \le Mr^j, \qquad j \ge 0,$$

for some constants $r \in (0,1)$ and $M \in (0,\infty)$. Then the equation $g - Sg = f$ has a solution $g \in \mathscr{H}$ if and only if

$$\lim_{n \to \infty} \frac{1}{n} \left\| \sum_0^n S^j f \right\|^2 = 0.$$

(iii) Call a function $f(x)$ in $\mathscr{H}_0$ smooth if the coefficients in the expansion

$$f(x) = \sum_{\substack{j = -\infty \\ j \ne 0}}^{\infty} a_j e^{2\pi i j x}$$

satisfy $\sum_{j=-\infty}^\infty |a_{(2j+1)2^n}|^2 \le Mr^n$, $n \ge 0$, for some constants $r \in (0,1)$ and $M \in (0,\infty)$. For $f(x)$ to be smooth, it is sufficient that it satisfy a Hölder condition of order $> \frac{1}{2}$ (Rochberg [1968]).

THEOREM. *If $f(x)$ is a smooth function in $\mathscr{H}_0$, then a necessary and sufficient condition for the existence of a $g(x)$ in $\mathscr{H}_0$ such that*

$$g(x) - g(2x) = f(x) \quad a.e. \text{ on } (-\infty, \infty)$$

is

$$\lim_{n \to \infty} \frac{1}{n} \int_0^1 \left| \sum_0^n f(2^j x) \right|^2 dx = 0.$$

When a solution $g(x)$ exists, it is unique and also smooth. See also Kac [1946].

9. A shift operator $S \in \mathscr{B}(\mathscr{H})$ of multiplicity 1 has no square root in $\mathscr{B}(\mathscr{H})$ (Halmos [1982]).

10. If C_1 is defined on $L^2(0,1)$ by

$$(C_1 f)(x) = x^{-1} \int_0^x f(t) \, dt, \qquad 0 < x < 1,$$

then $I - C_1^*$ is a shift operator of multiplicity 1 (de Branges [1961], Brown, Halmos, and Shields [1965]).

Notes

1.1   The terms "shift operator" and "unilateral shift operator" are synonymous in the literature. Bilateral shift operators are different objects (Halmos [1961], [1982]).

1.3   The Wold decomposition is due to von Neumann [1929a] and Wold [1938]. A Banach space version is given in Faulkner and Huneycutt [1978].

1.5   The concept of a universal model for Hilbert space operators is due to Rota [1959], [1960]. Rota's model is based on similarity rather than unitary equivalence. The model in Section 1.5 is a variant of Rota's model that was used in an unsuccessful attempt on the invariant subspace problem by de Branges and Rovnyak [1964], [1966a,b]. A version of the model for operators satisfying $\|T\| < 1$ was given in Rovnyak [1963]. Independently, Foiaş [1963] constructed a similar model by a different method. See also Sz.-Nagy and Foiaş [1964], [1970] and Helson [1964]. Other generalizations and refinements of the model are given in Agler [1982], Ball [1977], Frazho [1982], and Durszt and Sz.-Nagy [1983].

1.9   This result is due to Sz.-Nagy and Foiaş [1966], [1970] and, in different form, de Branges and Rovnyak [1966b].

1.12   Beurling [1949] obtained the representation of an invariant subspace for a shift operator of multiplicity 1, and Lax [1959] extended Beurling's result to shift operators of arbitrary finite multiplicity. Lax formulated his extension of Beurling's theorem as a characterization of the invariant subspaces of a semigroup of translation operators. However, using the connection between a semigroup of contraction operators and its cogenerator, one can see that Lax's theorem is equivalent to 1.12 in the case of a shift operator of finite multiplicity (Beals [1971], pp. 76–77). Halmos [1961] and, independently, Rovnyak [1962] gave direct proofs valid for shift operators of arbitrary multiplicity. Our proof of 1.12 is a simplified version of the proof given in Rosenblum and Rovnyak [1971].

It is a curiosity that the Beurling-Lax theorem is logically simpler for infinite multiplicity shifts than for finite multiplicity shifts. The reason for this is that the dimension inequality required for the proof (dim $Q\mathcal{H} \leq$ dim $\mathcal{K}$) is trivial in the infinite multiplicity case. Every proof uses some such inequality, and the first proofs of the inequality were awkward. This difficulty was overcome by I. Halperin, who showed that the key inequality can be obtained from the Counting Lemma 1.11.

1.13   The lifting theorem is due to Sarason [1967] in a special case and Sz.-Nagy and Foiaş [1968a,b], [1970] in general.

Additional Literature
Ball and Helton [1982a], [1983], Fillmore [1970], Helson [1964], Hoffman [1962], McEnnis [1979], [1982], Nikol'skiĭ [1980], Radjavi and Rosenthal [1973], and Sz.-Nagy and Foiaş [1970].

Chapter 2

Interpolation Theorems of the Pick-Nevanlinna
and Loewner Types

2.1 Introduction

For $\Omega = D$ or Π, let $B(\Omega)$ be the set of all functions $w(z)$ that are holomorphic and bounded by 1 on Ω. We are concerned with interpolation theorems for $B(\Omega)$—that is, characterizations of functions in $B(\Omega)$ in terms of data on subsets of Ω or $\partial\Omega$. When the data are prescribed in Ω, two classical theorems serve as prototypes: the Pick-Nevanlinna theorem and the Carathéodory-Fejér theorem (see Sections 2.4 and 2.5). The prototype for the situation in which data are prescribed on $\partial\Omega$ is Loewner's theorem (see 2.11). We use an operator method based on the lifting theorem 1.13 to prove both the classical theorems and some recent generalizations. We also sketch the theory of monotone operator functions in the Examples and Addenda at the end of the chapter.

2.2 Nudel'man's Problem

For any complex vector space $\mathscr{V}$, let $\mathscr{V}'$ be the space of all linear functionals on $\mathscr{V}$, and let $\mathscr{L}(\mathscr{V})$ be the space of all linear operators on $\mathscr{V}$ to $\mathscr{V}$. The value of a functional $x' \in \mathscr{V}'$ on a vector $x \in \mathscr{V}$ is written (x,x'). Each $A \in \mathscr{L}(\mathscr{V})$ induces an operator $A' \in \mathscr{L}(\mathscr{V}')$ such that

$$(Ax,x') = (x,A'x')$$

for all $x \in \mathscr{V}$ and $x' \in \mathscr{V}'$. We use no topology on $\mathscr{V}$, so questions concerning continuity do not arise.

PROBLEM.   *Given $A \in \mathscr{L}(\mathscr{V})$ and $b,c \in \mathscr{V}$, find conditions for the existence of $w \in B(\Omega)$, $\Omega = D$ or Π, such that*

$$b = w(A)c. \tag{2-1}$$

It is necessary to give meaning to (2-1). One way to do this is to choose a functional calculus in which $w(A)$ is defined. For example, $w(A)$ is defined by a standard calculus if $\mathscr{V}$ is finite dimensional and the eigenvalues of A lie in Ω (see Gantmacher [1959], vol. I, Hamburger and Grimshaw [1956], or MacDuffee [1946]). Even for finite-dimensional $\mathscr{V}$, Nudel'man's problem is quite interesting.

We adopt a different approach in which we give meaning to the vector equation (2-1) without ever defining an operator $w(A)$. This approach is consistent with standard functional calculi in special cases.

It is enough for our purposes to treat the case $\Omega = D$. Notation is as above.

2.3   Main Theorem

Let $A \in \mathcal{L}(\mathcal{V})$ and $b,c \in \mathcal{V}$ be given. Let $\mathcal{D} \subseteq \mathcal{V}'$ be a linear subspace such that $A'\mathcal{D} \subseteq \mathcal{D}$ and

$$\sum_0^\infty |(A^j c, x')|^2 < \infty$$

for every $x' \in \mathcal{D}$. The following are equivalent:

(i)   There exists $w(z) = \sum_0^\infty w_j z^j$ in $B(D)$ such that

$$(b, x') = \sum_0^\infty w_j (A^j c, x'), \qquad x' \in \mathcal{D}. \tag{2-2}$$

(ii)   For all $x' \in \mathcal{D}$,

$$\sum_0^\infty |(A^j b, x')|^2 \le \sum_0^\infty |(A^j c, x')|^2. \tag{2-3}$$

Proof.   Think of $H^2(D)$ as the space of power series $\sum_0^\infty a_j z^j$ with square summable coefficients. Let S be the shift operator

$$S: f(z) \to z f(z)$$

on $H^2(D)$.

Assume (ii). We apply the lifting theorem in 1.13 with $\mathcal{G}_j = H^2(D)$ and $S_j = S, j = 1,2$. For each $x' \in \mathcal{D}$, $\sum_0^\infty (A^j c, x') z^j$ is in $H^2(D)$, and

$$S^* \left\{ \sum_0^\infty (A^j c, x') z^j \right\} = \sum_0^\infty (A^{j+1} c, x') z^j = \sum_0^\infty (A^j c, A' x') z^j. \tag{2-4}$$

Let $\mathcal{H}_1$ be the closure in $H^2(D)$ of all series $\sum_0^\infty (A^j c, x') z^j$, where $x' \in \mathcal{D}$. Since $A'\mathcal{D} \subseteq \mathcal{D}$, it follows from (2-4) that $\mathcal{H}_1$ is invariant under S^*. Let $\mathcal{H}_2 = H^2(D)$. Let $T_j = S^*|\mathcal{H}_j, j = 1,2$. By (ii) there is a unique operator $X \in \mathcal{B}(\mathcal{H}_1, \mathcal{H}_2)$ such that $\|X\| \le 1$ and for each $x' \in \mathcal{D}$,

$$X \left\{ \sum_0^\infty (A^j c, x') z^j \right\} = \sum_0^\infty (A^j b, x') z^j.$$

For each $x' \in \mathcal{D}$,

$$XS^* \left\{ \sum_0^\infty (A^j c, x') z^j \right\} = \sum_0^\infty (A^j b, A' x') z^j$$

$$= S^* X \left\{ \sum_0^\infty (A^j c, x') z^j \right\}.$$

Thus $X T_1 = T_2 X$, and the hypotheses of the lifting theorem are satisfied. By the lifting theorem, $X = Y|\mathcal{H}_1$ for some operator Y on $H^2(D)$ such that $YS^* = S^*Y$ and $\|Y\| = \|X\| \le 1$. Applying Theorem B (1.15) in the scalar case, we obtain

$$Y^* f = \tilde{w} f, \qquad f \in H^2(D),$$

for some function $\tilde{w}(z) = \sum_0^\infty \bar{w}_j z^j$ in $B(D)$. For any $x' \in \mathscr{D}$,

$$
\begin{aligned}
(b,x') &= \left\langle \sum_0^\infty (A^j b, x') z^j, 1 \right\rangle_2 \\
&= \left\langle X \left\{ \sum_0^\infty (A^j c, x') z^j \right\}, 1 \right\rangle_2 \\
&= \left\langle \sum_0^\infty (A^j c, x') z^j, Y^*\{1\} \right\rangle_2 \\
&= \left\langle \sum_0^\infty (A^j c, x') z^j, \sum_0^\infty \bar{w}_j z^j \right\rangle_2 \\
&= \sum_0^\infty (A^j c, x') w_j.
\end{aligned}
$$

Thus $w(z) = \sum_0^\infty w_j z^j$ is in $B(D)$ and (i) holds.

Conversely, assume (i). Let Y be the operator on $H^2(D)$ such that Y^* is multiplication by $\tilde{w}(z) = \sum_0^\infty \bar{w}_j z^j$. Then $\|Y\| \leq 1$. By (2-2), for each $x' \in \mathscr{D}$ and $k = 0,1,2,\ldots,$

$$
(A^k b, x') = (b, A'^k x') = \sum_0^\infty (A^j c, A'^k x') w_j = \sum_0^\infty (A^{j+k} c, x') w_j.
$$

Hence if $Y\{\sum_0^\infty (A^j c, x') z^j\} = \sum_0^\infty g_k z^k$, then

$$
\begin{aligned}
g_k &= \left\langle Y \left\{ \sum_0^\infty (A^j c, x') z^j \right\}, S^k\{1\} \right\rangle_2 \\
&= \left\langle \sum_0^\infty (A^j c, x') z^j, Y^* S^k\{1\} \right\rangle_2 \\
&= \left\langle \sum_0^\infty (A^j c, x') z^j, \sum_0^\infty \bar{w}_j z^{j+k} \right\rangle_2 \\
&= \sum_0^\infty (A^{l+k} c, x') w_l \\
&= (A^k b, x').
\end{aligned}
$$

We have shown that for all $x' \in \mathscr{D}$,

$$
Y\left\{ \sum_0^\infty (A^j c, x') z^j \right\} = \sum_0^\infty (A^j b, x') z^j.
$$

Since $\|Y\| \leq 1$, (2-3) holds and (i) follows. ∎

REMARK.   The lifting theorem is not needed in the proof if $\mathscr{H}_1 = H^2(D)$. This case occurs in interesting and nontrivial situations such as the applications of 2.7.

2.4   The Pick and Pick-Nevanlinna Theorems

Special cases in Section 2.3 yield classical interpolation theorems.

PICK'S THEOREM.   *Let* $z_1,\ldots,z_n \in D$ *and* $w_1,\ldots,w_n \in \mathbf{C}$. *There exists* $w \in B(D)$ *such that* $w(z_j) = w_j, j = 1,\ldots,n$, *if and only if*

$$\sum_{j,k=1}^{n} \frac{1 - w_j\bar{w}_k}{1 - z_j\bar{z}_k} a_j\bar{a}_k \geq 0 \tag{2-5}$$

for all $a_1,\ldots,a_n \in \mathbf{C}$.

Proof.   In Section 2.3 let $\mathscr{V} = \mathbf{C}^n$ be viewed as a space of column vectors. Let $A = \operatorname{diag}\{z_1,\ldots,z_n\}$, $b = [w_1,\ldots,w_n]^t$, $c = [1,\ldots,1]^t$, and $\mathscr{D} = \mathscr{V}'$. The t denotes matrix transpose. The hypotheses of the theorem are satisfied. If we represent linear functionals on $\mathbf{C}^n$ in the usual way, it is easy to see that condition (i) of Section 2.3 asserts the existence of $w \in B(D)$ such that

$$\sum_{p=1}^{n} \bar{a}_p w_p = \sum_{p=1}^{n} \bar{a}_p w(z_p)$$

for all $a_1,\ldots,a_n \in \mathbf{C}$, that is, $w_p = w(z_p)$, $p = 1,\ldots,n$. Condition (ii) of 2.3 holds if and only if

$$\sum_{j=0}^{\infty} \left| \sum_{p=1}^{n} \bar{a}_p w_p z_p^j \right|^2 \leq \sum_{j=0}^{\infty} \left| \sum_{p=1}^{n} \bar{a}_p z_p^j \right|^2.$$

Expanding and summing first with respect to j, we see that this inequality is the same as (2-5). Thus the result follows from 2.3.   ∎

PICK-NEVANLINNA THEOREM.   *Let* $\{z_j\}_{j \in J} \subseteq D$ *and* $\{w_j\}_{j \in J} \subseteq \mathbf{C}$. *There exists* $w \in B(D)$ *such that* $w(z_j) = w_j$ *for all* $j \in J$ *if and only if*

$$\sum_{j,k \in J} \frac{1 - w_j\bar{w}_k}{1 - z_j\bar{z}_k} a_j\bar{a}_k \geq 0 \tag{2-6}$$

for all $\{a_j\}_{j \in J} \subseteq \mathbf{C}$ *such that* $\{ j: a_j \neq 0\}$ *is finite.*

Usually this is deduced from Pick's theorem by a normal families argument. We prove it directly using 2.3.

Proof.   Let $\mathscr{V} = \mathbf{C}^{\operatorname{card} J}$ be the space of all indexed sets $x = \{x_j\}_{j \in J}$ in $\mathbf{C}$ with coordinatewise addition and scalar multiplication. Define $A \in \mathscr{L}(\mathscr{V})$ and $b, c \in \mathscr{V}$ by

$$A: \{x_j\}_{j \in J} \to \{z_j x_j\}_{j \in J},$$

$b = \{w_j\}_{j \in J}$, and $c = \{c_j\}_{j \in J}$, where $c_j = 1$, $j \in J$. Let $\mathscr{D}$ be the set of all linear functionals x' on $\mathscr{V}$ of the form

$$(x,x') = \sum_{j \in J} \bar{a}_j x_j,$$

where $\{a_j\}_{j \in J} \subseteq \mathbf{C}$ and $\{j : a_j \neq 0\}$ is finite. The hypotheses of 2.3 are satisfied, and the result follows by a routine generalization of the proof of Pick's theorem.

∎

2.5 Carathéodory-Fejér Theorem

For $a_0, \ldots, a_n \in \mathbf{C}$, set

$$
T(a_0, \ldots, a_n) = \begin{bmatrix}
a_0 & 0 & 0 & \cdots & 0 \\
a_1 & a_0 & 0 & \cdots & 0 \\
a_2 & a_1 & a_0 & \cdots & 0 \\
& & \cdots & & \\
a_n & a_{n-1} & a_{n-2} & \cdots & a_0
\end{bmatrix}_{(n+1) \times (n+1)}.
\tag{2-7}
$$

The norm $\|M\|$ of an $(n + 1) \times (n + 1)$ matrix M is its norm as an operator on $\mathbf{C}^{n+1}$ in the usual inner product.

CARATHÉODORY-FEJÉR THEOREM.   *Given $b_0, \ldots, b_n \in \mathbf{C}$, there exists $w \in B(D)$ such that*

$$w(z) = b_0 + b_1 z + \cdots + b_n z^n + \text{higher powers}$$

if and only if

$$\|T(b_0, \ldots, b_n)\| \leq 1.$$

With no extra effort we obtain a more general result.

THEOREM.   *Given $b_0, \ldots, b_n, c_0, \ldots, c_n \in \mathbf{C}$, there exists $w(z) = \sum_0^\infty w_j z^j$ in $B(D)$ such that*

$$
\begin{cases}
b_0 = c_0 w_0 \\
b_1 = c_0 w_1 + c_1 w_0 \\
\quad \vdots \\
b_n = c_0 w_n + c_1 w_{n-1} + \cdots + c_n w_0
\end{cases}
\tag{2-8}
$$

if and only if

$$T(b_0, \ldots, b_n) T(b_0, \ldots, b_n)^* \leq T(c_0, \ldots, c_n) T(c_0, \ldots, c_n)^*.
\tag{2-9}$$

Proof.   In 2.3, let $\mathcal{V} = \mathbf{C}^{n+1}$, $A = T(0,1,0,\ldots,0)$, $b = [b_0, \ldots, b_n]^t$, $c = [c_0, \ldots, c_n]^t$, and $\mathcal{D} = \mathcal{V}'$. Since $A^j = 0$ for $j > n$, condition (i) of 2.3 asserts the existence of $w(z) = \sum_0^\infty w_j z^j$ in $B(D)$ such that for all $x' \in \mathcal{V}'$,

$$(b, x') = w_0(c, x') + w_1(Ac, x') + \cdots + w_n(A^n c, x'),$$

that is, such that (2-8) holds. Any $x' \in \mathcal{V}'$ can be represented as an inner product $(x, x') = \langle x, a \rangle = a^* x$ for some $a = [a_0, \ldots, a_n]^t \in \mathbf{C}^{n+1}$. Then

$$
\sum_0^\infty |(A^j b, x')|^2 = |a^* b|^2 + |a^* A b|^2 + \cdots + |a^* A^n b|^2
$$

$$
= |\bar{b}_0 a_0 + \cdots + \bar{b}_n a_n|^2 + |\bar{b}_0 a_1 + \cdots + \bar{b}_{n-1} a_n|^2 + \cdots + |\bar{b}_0 a_n|^2
$$

$$
= \|T(b_0, \ldots, b_n)^* a\|^2.
$$

Similarly with b replaced by c. Thus condition (ii) of 2.3 asserts that for all $a \in \mathbf{C}^{n+1}$,

$$\|T(b_0,\ldots,b_n)^*a\|^2 \leq \|T(c_0,\ldots,c_n)^*a\|^2,$$

and (2-9) holds. The result thus follows from 2.3. ∎

2.6 Simultaneous Generalization of the Pick-Nevanlinna and Carathéodory-Fejér Theorems

For $\alpha,\beta \in D$, set

$$\frac{1}{1 - (\alpha + s)(\bar{\beta} + \bar{t})} = \sum_{p,q=0}^{\infty} K_{pq}(\alpha,\beta)s^p\bar{t}^q \tag{2-10}$$

for $|s|$, $|t|$ sufficiently small. Differentiation of the identity

$$\frac{1}{1 - \alpha\bar{\beta}} = \int_{\Gamma} \frac{d\sigma}{(1 - \alpha e^{i\theta})(1 - \bar{\beta} e^{-i\theta})}$$

yields

$$K_{pq}(\alpha,\beta) = \frac{1}{p!\,q!}\left(\frac{\partial}{\partial\alpha}\right)^p\left(\frac{\partial}{\partial\bar{\beta}}\right)^q \frac{1}{1 - \alpha\bar{\beta}}$$

$$= \int_{\Gamma} \frac{e^{i(p-q)\theta}}{(1 - \alpha e^{i\theta})^{p+1}(1 - \bar{\beta}e^{-i\theta})^{q+1}}\,d\sigma \tag{2-11}$$

for all $p,q = 0,1,2,\ldots$.

For simplicity, consider first the Pick and Carathéodory-Fejér theorems. Define $T(a_0,\ldots,a_n)$ for $a_0,\ldots,a_n \in \mathbf{C}$ as in (2-7).

THEOREM A.   Let $Z = \{z_j\}_{j=1}^n \subseteq D$. For each $j = 1,\ldots,n$, let $b_{j0},\ldots,b_{j,r(j)}$ be given complex numbers. There exists $w \in B(D)$ such that for each $j = 1,\ldots,n$,

$$w(z) = b_{j0} + b_{j1}(z - z_j) + \cdots + b_{j,r(j)}(z - z_j)^{r(j)} + \text{higher powers}$$

if and only if

$$T(b)P(Z)T(b)^* \leq P(Z),$$

where $T(b) = [T_{jk}(b)]_{j,k=1}^n$ and $P(Z) = [P_{jk}(Z)]_{j,k=1}^n$ are block matrices defined as follows: For each $j,k = 1,\ldots,n$, $T_{jk}(b)$ and $P_{jk}(Z)$ have order $(r(j) + 1) \times (r(k) + 1)$, and

$$T_{jk}(b) = T(b_{j0},\ldots,b_{j,r(j)}) \qquad \text{if } j = k,$$

$$T_{jk}(b) = 0 \qquad \text{if } j \neq k,$$

$$P_{jk}(Z) = [K_{pq}(z_j,z_k)]_{\substack{p=0,\ldots,r(j) \\ q=0,\ldots,r(k)}}.$$

This result includes as special cases both the Pick theorem in Section 2.4 and the Carathéodory-Fejér theorem in Section 2.5.

THEOREM B.   Let $Z = \{z_j\}_{j \in J} \subseteq D$, and for each $j \in J$ let $b_{j0}, \ldots, b_{j,r(j)}$ and $c_{j0}, \ldots, c_{j,r(j)}$ be given complex numbers. There exists $w \in B(D)$ such that for each $j \in J$ the coefficients in the expansion

$$w(z) = w_{j0} + w_{j1}(z - z_j) + w_{j2}(z - z_j)^2 + \cdots \tag{2-12}$$

satisfy

$$\begin{cases} b_{j0} = c_{j0}w_{j0} \\ b_{j1} = c_{j0}w_{j1} + c_{j1}w_{j0} \\ \quad \vdots \\ b_{j,r(j)} = c_{j0}w_{j,r(j)} + c_{j1}w_{j,r(j)-1} + \cdots + c_{j,r(j)}w_{j0} \end{cases} \tag{2-13}$$

if and only if

$$T(b)P(Z)T(b)^* \le T(c)P(Z)T(c)^*. \tag{2-14}$$

The notation in (2-14) is similar to that used in Theorem A: Define $T(b) = \{T_{jk}(b)\}_{j,k \in J}$ and $P(Z) = \{P_{jk}(Z)\}_{j,k \in J}$ as in Theorem A, except that the index set J is used in place of $\{1, \ldots, n\}$. Define $T(c)$ similarly by replacing all of the b's by c's. The meaning of (2-14) is that

$$\sum_{j,k \in J} a_j^* T_{jj}(b) P_{jk}(Z) T_{kk}(b)^* a_k$$

$$\le \sum_{j,k \in J} a_j^* T_{jj}(c) P_{jk}(Z) T_{kk}(c)^* a_k \tag{2-15}$$

whenever $a_j \in \mathbf{C}^{r(j)+1}, j \in J$, and $\{j : a_j \ne 0\}$ is finite.

Theorem B includes Theorem A. It is sufficient (and no more difficult) to prove Theorem B.

Proof of Theorem B.   Let $\mathscr{V}$ be the set of all indexed sets $x = \{x_j\}_{j \in J}$, where $x_j \in \mathbf{C}^{r(j)+1}, j \in J$, with linear operations defined coordinatewise. Let

$$A : \{x_j\}_{j \in J} \to \{A_j x_j\}_{j \in J},$$

where

$$A_j = \begin{bmatrix} z_j & & & & \\ 1 & z_j & & \text{\Large 0} & \\ & 1 & & & \\ & & \ddots & & \\ & & & \ddots & \\ \text{\Large 0} & & & 1 & z_j \end{bmatrix}_{(r(j)+1) \times (r(j)+1)}, \qquad j \in J.$$

Let $b = \{b_j\}_{j \in J}, c = \{c_j\}_{j \in J}$, where for all $j \in J$,

$$b_j = [b_{j0}, \ldots, b_{j,r(j)}]^t \qquad \text{and} \qquad c_j = [c_{j0}, \ldots, c_{j,r(j)}]^t.$$

Let $\mathscr{D}$ be the set of all linear functionals x' on $\mathscr{V}$ of the form

$$(x,x') = \sum_{j \in J} a_j^* x_j, \tag{2-16}$$

where $a_j \in \mathbf{C}^{r(j)+1}, j \in J$, and $\{j: a_j \neq 0\}$ is finite.

It is convenient to introduce a functional calculus for A. For any holomorphic function $f(z)$ on D, define $f(A)$ on $\mathscr{V}$ by

$$f(A): \{x_j\}_{j \in J} \rightarrow \{f(A_j)x_j\}_{j \in J},$$

where for each j, $f(A_j)$ is defined by the standard matrix calculus. The main fact concerning the matrix calculus that we need is this: for any square matrix M with eigenvalues in D, if

$$f(z) = f_0 + f_1 z + f_2 z^2 + \cdots, \text{ then } f(M) = f_0 I + f_1 M + f_2 M^2 + \cdots.$$

Condition (i) of 2.3 asserts the existence of $w \in B(D)$ such that $b = w(A)c$—that is, $b_j = w(A_j)c_j$ for all $j \in J$—that is, the coefficients in (2-12) satisfy (2-13) for all $j \in J$.

We interpret condition (ii) in 2.3. Let $u = e^{i\theta}$, and let $x' \in \mathscr{D}$ be given by (2-16). Then

$$\sum_0^\infty |(A^p b, x')|^2 = \int_\Gamma \left| \sum_0^\infty (A^p b, x') u^p \right|^2 d\sigma(u),$$

$$= \int_\Gamma |((I - Au)^{-1}b, x')|^2 \, d\sigma(u)$$

$$= \int_\Gamma \left| \sum_{j \in J} a_j^* (I_j - A_j u)^{-1} b_j \right|^2 d\sigma(u), \tag{2-17}$$

where I is the identity operator on $\mathscr{V}$ and I_j is the identity matrix on $\mathbf{C}^{r(j)+1}, j \in J$. For any $j \in J$ and $u \in \Gamma$, set

$$h_j(u) = [1/(1 - z_j u), u/(1 - z_j u)^2, \ldots, u^{r(j)}/(1 - z_j u)^{r(j)+1}]^t.$$

Then

$$(I_j - A_j u) T_{jj}(b) h_j(u) = b_j,$$

and so

$$(I_j - A_j u)^{-1} b_j = T_{jj}(b) h_j(u).$$

Hence by (2-17)

$$\sum_0^\infty |(A^p b, x')|^2 = \int_\Gamma \left| \sum_{j \in J} a_j^* T_{jj}(b) h_j(u) \right|^2 d\sigma(u)$$

$$= \int_\Gamma \sum_{j,k \in J} a_j^* T_{jj}(b) h_j(u) h_k(u)^* T_{kk}(b)^* a_k \, d\sigma(u)$$

$$= \sum_{j,k \in J} a_j^* T_{jj}(b) P_{jk}(Z) T_{kk}(b)^* a_k.$$

For the last equality we used the identity

$$P_{jk}(Z) = \int_\Gamma h_j(u)h_k(u)^* \, d\sigma(u), \qquad j,k \in J,$$

which follows from (2-11). Similarly,

$$\sum_0^\infty |(A^p c, x')|^2 = \sum_{j,k \in J} a_j^* T_{jj}(c)P_{jk}(Z)T_{kk}(c)^* a_k.$$

The inequality (2-3) is thus the same as (2-15), and so the result follows from 2.3.

∎

2.7   Restrictions of Boundary Functions

We characterize the restrictions of boundary functions of functions in $B(\Omega)$, $\Omega = D$ or Π, to an arbitrary Borel subset Δ of $\partial\Omega$.

To shorten formulas in the disk case, we write $u = e^{i\theta}$, $v = e^{i\psi}$ for typical points on $\Gamma = \partial D$. Measure theoretic notions are relative to normalized Lebesgue measure σ on Γ.

THEOREM A.   *Let b,c be measurable complex valued functions on a Borel set $\Delta \subseteq \Gamma$. There exists $w \in B(D)$ such that*

$$b(u) = w(u)c(u) \qquad \sigma\text{-a.e. on } \Delta \qquad\qquad (2\text{-}18)$$

if and only if

$$\lim_{r\uparrow 1} \int_\Delta \int_\Delta \frac{c(u)\bar{c}(v) - b(u)\bar{b}(v)}{1 - r^2 u\bar{v}} \phi(u)\bar{\phi}(v) \, d\sigma(u) \, d\sigma(v) \geq 0 \qquad\qquad (2\text{-}19)$$

for every measurable complex valued function ϕ on Δ such that $b\phi, c\phi \in L^2(\Delta)$.

The proof shows that the limit in (2-19) exists as a real number for each ϕ such that $b\phi, c\phi \in L^2(\Delta)$, whether or not (2-19) holds.

Proof.   Let $\mathcal{V}$ be the space of complex valued functions on Δ of the form $f = pb + qc$, where p,q are polynomials. Thus $b,c \in \mathcal{V}$. Define

$$A: f(u) \to uf(u)$$

on $\mathcal{V}$. Let $\mathcal{D}$ be the set of linear functionals x' on $\mathcal{V}$ of the form

$$(f,x') = \int_\Delta f\phi \, d\sigma, \qquad f \in \mathcal{V}, \qquad\qquad (2\text{-}20)$$

where ϕ is a measurable function such that $b\phi, c\phi \in L^2(\Delta)$. Then $A'\mathcal{D} \subseteq \mathcal{D}$. For every functional (2-20), $\sum_0^\infty |(A^j c, x')|^2 < \infty$ because the Fourier coefficients of a square summable function are square summable. Thus the hypotheses of 2.3 are satisfied.

Condition (i) of 2.3 holds if and only if there exists $w(z) = \sum_0^\infty w_j z^j$ in $B(D)$ such that

$$\int_\Delta b(u)\phi(u)\,d\sigma(u) = \sum_0^\infty w_j \int_\Delta u^j c(u)\phi(u)\,d\sigma(u)$$

$$= \lim_{r\uparrow 1} \sum_0^\infty r^j w_j \int_\Delta u^j c(u)\phi(u)\,d\sigma(u)$$

$$= \lim_{r\uparrow 1} \int_\Delta w(ru)c(u)\phi(u)\,d\sigma(u)$$

$$= \int_\Delta w(u)c(u)\phi(u)\,d\sigma(u)$$

for all ϕ such that $b\phi, c\phi \in L^2(\Delta)$, that is, (2-18) holds.

Condition (ii) of 2.3 asserts that for all ϕ such that $b\phi, c\phi \in L^2(\Delta)$,

$$\sum_0^\infty \left| \int_\Delta u^j b\phi\,d\sigma \right|^2 \le \sum_0^\infty \left| \int_\Delta u^j c\phi\,d\sigma \right|^2. \tag{2-21}$$

Now

$$\sum_0^\infty \left| \int_\Delta u^j b\phi\,d\sigma \right|^2$$

$$= \lim_{r\uparrow 1} \int_\Gamma \left| \sum_0^\infty \left(\int_\Delta u^j b\phi\,d\sigma \right) r^j t^j \right|^2 d\sigma(t)$$

$$= \lim_{r\uparrow 1} \int_\Gamma \left| \int_\Delta \frac{b(u)\phi(u)}{1 - rut}\,d\sigma(u) \right|^2 d\sigma(t)$$

$$= \lim_{\cdot r\uparrow 1} \int_\Delta \int_\Delta \int_\Gamma \frac{d\sigma(t)}{(1 - rut)(1 - r\bar{v}\bar{t})} b(u)\bar{b}(v)\phi(u)\bar{\phi}(v)\,d\sigma(u)\,d\sigma(v)$$

$$= \lim_{r\uparrow 1} \int_\Delta \int_\Delta \frac{b(u)\bar{b}(v)}{1 - r^2 u\bar{v}} \phi(u)\bar{\phi}(v)\,d\sigma(u)\,d\sigma(v).$$

The change in the order of integration is justified by Fubini's theorem. A similar identity holds with b replaced by c. Hence (2-21) is the same as (2-19), and the result follows from 2.3.   ∎

THEOREM B.   *Let b,c be measurable complex valued functions on a Borel subset Δ of $(-\infty,\infty)$. There exists $w \in B(\Pi)$ such that*

$$b(x) = w(x)c(x) \qquad a.e. \text{ on } \Delta \tag{2-22}$$

if and only if

$$\lim_{\varepsilon \downarrow 0} i \int\limits_{\Delta} \int\limits_{\Delta} \frac{c(s)\bar{c}(t) - b(s)\bar{b}(t)}{s - t + i\varepsilon} \phi(s)\bar{\phi}(t) \, ds \, dt \geq 0 \qquad (2\text{-}23)$$

for all measurable complex valued functions ϕ on Δ such that $b\phi$, $c\phi \in L^2(\Delta)$.

It would be awkward to deduce this from Theorem A by a change of variables because of the limit interchanges. We give a direct proof using 2.3 instead. As in Theorem A, the limit in (2-23) exists as a real number for every ϕ such that $b\phi$, $c\phi \in L^2(\Delta)$, whether or not (2-23) holds.

Proof. In 2.3, let A be multiplication by $(x - i)/(x + i)$ on the space $\mathscr{V}$ of all functions on Δ of the form

$$f(x) = p\left(\frac{x - i}{x + i}\right)b(x) + q\left(\frac{x - i}{x + i}\right)c(x)$$

where p, q are polynomials. In particular, $b, c \in \mathscr{V}$. Let $\mathscr{D}$ be the set of linear functionals x' on $\mathscr{V}$ of the form

$$(f, x') = \int\limits_{\Delta} (t + i)^{-1} f\phi \, dt, \qquad f \in \mathscr{V}.$$

The hypotheses of 2.3 are readily verified.

Suppose condition (i) of 2.3 holds. Then there exists $h(z) = \sum_0^\infty h_j z^j$ in $B(D)$ such that for each ϕ with $b\phi, c\phi \in L^2(\Delta)$,

$$\int\limits_{\Delta} \frac{b(t)\phi(t)}{t + i} \, dt = \sum_0^\infty h_j \int\limits_{\Delta} \left(\frac{t - i}{t + i}\right)^j \frac{c(t)\phi(t)}{t + i} \, dt$$

$$= \lim_{r \uparrow 1} \sum_0^\infty r^j h_j \int\limits_{\Delta} \left(\frac{t - i}{t + i}\right)^j \frac{c(t)\phi(t)}{t + i} \, dt$$

$$= \lim_{r \uparrow 1} \int\limits_{\Delta} h\left(r\frac{t - i}{t + i}\right) \frac{c(t)\phi(t)}{t + i} \, dt$$

$$= \int\limits_{\Delta} h\left(\frac{t - i}{t + i}\right) \frac{c(t)\phi(t)}{t + i} \, dt.$$

Hence

$$b(x) = h\left(\frac{x - i}{x + i}\right)c(x) \qquad \text{a.e. on } \Delta.$$

Define

$$w(z) = h\left(\frac{z - i}{z + i}\right), \qquad z \in \Pi.$$

Then $w \in B(\Pi)$ and (2-22) holds. Reversing the steps, we see that if there exists $w \in B(\Pi)$ such that (2-22) holds, then condition (i) of 2.3 holds.

We interpret condition (ii) of 2.3. Since

$$\left\{ \frac{\pi^{-1/2}}{z+i} \left(\frac{z-i}{z+i} \right)^j \right\}_0^\infty$$

is an orthonormal basis for $H^2(\Pi)$, if $\{a_j\}_0^\infty \in l^2$ and

$$F(z) = \sum_0^\infty a_j \frac{\pi^{-1/2}}{z+i} \left(\frac{z-i}{z+i} \right)^j, \qquad z \in \Pi, \tag{2-24}$$

then

$$\sum_0^\infty |a_j|^2 = \lim_{y \downarrow 0} \int_{-\infty}^{\infty} |F(x+iy)|^2 \, dx. \tag{2-25}$$

Condition (ii) of 2.3 holds if and only if

$$\sum_0^\infty \left| \int_\Delta \left(\frac{t-i}{t+i} \right)^j \frac{b(t)\phi(t)}{t+i} \, dt \right|^2 \leq \sum_0^\infty \left| \int_\Delta \left(\frac{t-i}{t+i} \right)^j \frac{c(t)\phi(t)}{t+i} \, dt \right|^2 \tag{2-26}$$

for all ϕ such that $b\phi, c\phi \in L^2(\Delta)$. Since

$$\sum_0^\infty \int_\Delta \left(\frac{t-i}{t+i} \right)^j \frac{b(t)\phi(t)}{t+i} \, dt \, \frac{\pi^{-1/2}}{z+i} \left(\frac{z-i}{z+i} \right)^j = \frac{1}{2i\pi^{1/2}} \int_\Delta \frac{b(t)\phi(t)}{t+x+iy} \, dt$$

for all $z = x + iy \in \Pi$, we have by (2-24) and (2-25),

$$\sum_0^\infty \left| \int_\Delta \left(\frac{t-i}{t+i} \right)^j \frac{b(t)\phi(t)}{t+i} \, dt \right|^2$$

$$= \lim_{y \downarrow 0} \int_{-\infty}^{\infty} \left| \frac{1}{2i\pi^{1/2}} \int_\Delta \frac{b(t)\phi(t)}{t+x+iy} \, dt \right|^2 dx$$

$$= \lim_{y \downarrow 0} \frac{1}{4\pi} \int_\Delta \int_\Delta \int_{-\infty}^{\infty} \frac{dx}{(s+x+iy)(t+x-iy)} b(s)\phi(s)\bar{b}(t)\bar{\phi}(t) \, ds \, dt$$

$$= \lim_{y \downarrow 0} \frac{i}{2} \int_\Delta \int_\Delta \frac{b(s)\bar{b}(t)}{s-t+2iy} \phi(s)\bar{\phi}(t) \, ds \, dt \tag{2-27}$$

for all ϕ as above. At the next to last stage, the inside integral $\int_{-\infty}^{\infty}$ is evaluated using Cauchy's formula for the upper half-plane (Duren [1970], p. 195). The change in order of integration requires justification. If Δ is bounded, this follows easily from Fubini's theorem. To reduce to this case, it is sufficient to show that for any fixed $y > 0$,

$$\lim_{n \to \infty} \int\limits_{\Delta_n} \frac{b(t)\phi(t)}{t + x + iy} dt = \int\limits_{\Delta} \frac{b(t)\phi(t)}{t + x + iy} dt$$

in the metric of $L^2(-\infty,\infty)$, where $\Delta_n = \Delta \cap (-n,n)$, $n = 1,2,\ldots$. This follows from the inequality

$$\int\limits_{-\infty}^{\infty} \left| \frac{1}{2\pi i} \int\limits_{\Delta \backslash \Delta_n} \frac{b(t)\phi(t)}{t + x + iy} dt \right|^2 dx \leq \int\limits_{\Delta \backslash \Delta_n} |b\phi|^2 \, dt,$$

which can be proved, for example, using Fourier transforms. The change in order of integration is thus justified. Moreover, (2-27) holds with b replaced by c. Therefore (2-26) is equivalent to (2-23), and the result follows from 2.3. ∎

2.8   Pick Class

We define the *Pick class* $\mathscr{P}$ as the set of holomorphic functions f on Π such that

$$\operatorname{Im} f(z) \geq 0, \qquad z \in \Pi.$$

There is a one-to-one correspondence between $\mathscr{P}$ and $B(\Pi)\backslash\{1\}$ ("1" is the function identically 1 on Π): If $w \in B(\Pi)\backslash\{1\}$, then

$$f = i(1 + w)/(1 - w) \tag{2-28}$$

is in $\mathscr{P}$, and every f in $\mathscr{P}$ has such a representation. An equivalent form of (2-28) is

$$w = (f - i)/(f + i). \tag{2-29}$$

Each f in the Pick class $\mathscr{P}$ that is not identically zero is an outer function (Duren [1970], p.51). In particular, each f in $\mathscr{P}$ has a nontangential boundary function defined a.e. on $(-\infty,\infty)$.

2.9   Generalized Loewner Theorem

Let $f_0(x)$ be a measurable complex valued function on a Borel subset Δ of $(-\infty,\infty)$. There exists $f \in \mathscr{P}$ such that

$$f(x) = f_0(x) \qquad \text{a.e. on } \Delta$$

if and only if

$$\lim_{\varepsilon \downarrow 0} \int\limits_{\Delta}\int\limits_{\Delta} \frac{f_0(s) - \bar{f}_0(t)}{s - t + i\varepsilon} \phi(s)\bar{\phi}(t) \, ds \, dt \geq 0 \tag{2-30}$$

whenever ϕ, $f_0\phi \in L^2(\Delta)$.

Proof.   Apply Theorem B (Section 2.7), with $b = f_0 - i$, $c = f_0 + i$, and use the correspondence (2-28), (2-29) between $\mathscr{P}$ and $B(\Pi)\backslash\{1\}$.   ∎

2.10   Hilbert Transforms

The L^2 theory of Hilbert transforms is sufficient for our purposes. Although this is well known, we include statements of the principal results for the convenience of the reader and later reference.

If $\phi(x) \in L^2(-\infty,\infty)$, its Hilbert transform is defined by

$$(H\phi)(x) = PV\frac{1}{\pi} \int\limits_{-\infty}^{\infty} \frac{\phi(t)}{t - x} dt,$$

where PV indicates a Cauchy principal value integral:

$$PV\frac{1}{\pi} \int\limits_{-\infty}^{\infty} = \lim_{\varepsilon \downarrow 0} \frac{1}{\pi} \int\limits_{|t-x| > \varepsilon} . \tag{2-31}$$

THEOREM A.   *If $\phi(x) \in L^2(-\infty,\infty)$, then the limit in (2-31) exists a.e. and in the metric of $L^2(-\infty,\infty)$. If*

$$\Phi(z) = \frac{1}{2\pi i} \int\limits_{-\infty}^{\infty} \frac{\phi(t)}{t - z} dt, \qquad z \neq \bar{z},$$

then

$$\Phi(x + i0) - \Phi(x - i0) = \phi(x),$$

$$\Phi(x + i0) + \Phi(x - i0) = -i(H\phi)(x)$$

a.e. on $(-\infty, \infty)$, where $\Phi(x \pm i0) = \lim_{y \downarrow 0} \Phi(x \pm iy)$ whenever the limit exists.

THEOREM B.   *The operator H on $L^2(-\infty,\infty)$ coincides with iI on $H^2(R)$ and $-iI$ on $(H^2(R))^- = \{\bar{\phi} : \phi \in H^2(R)\}$. In particular, for any $\phi,\psi \in L^2(-\infty, \infty)$,*

$$\langle H\phi,\psi\rangle_2 = -\langle \phi,H\psi\rangle_2,$$

$$\|H\phi\|_2 = \|\phi\|_2 \quad and \quad H^2\phi = -\phi.$$

Moreover, if $\phi \in H^2(R)$ and $\phi = u + iv$, where u,v are real valued, then

$$Hv = u \quad and \quad Hu = -v.$$

CONVOLUTION THEOREM.   *If $\phi,\psi \in L^2(-\infty,\infty) \cap L^\infty(-\infty,\infty)$, then*

$$H(\phi H\psi + (H\phi)\psi) = (H\phi)(H\psi) - \phi\psi$$

a.e. on $(-\infty,\infty)$.

See Butzer and Nessel [1971], Titchmarsh [1948], Tricomi [1957], and Zygmund [1968].

2.11   Equivalent Form of the Generalized Loewner Theorem

Let Δ be a fixed Borel subset of $(-\infty,\infty)$. If $\phi \in L^2(\Delta)$, set

$$(H_\Delta \phi)(x) = PV \frac{1}{\pi} \int_\Delta \frac{\phi(t)}{t-x}\, dt \qquad \text{a.e. on } \Delta. \tag{2-32}$$

This defines a bounded linear operator on $L^2(\Delta)$ with adjoint $H_\Delta^* = -H_\Delta$.

THEOREM. *Let $f_0(x)$ be a measurable complex valued function on Δ. There exists $f \in \mathscr{P}$ such that*

$$f(x) = f_0(x) \qquad \text{a.e. on } \Delta$$

if and only if

$$\mathrm{Re}\langle (H_\Delta - iI)f_0\phi,\phi\rangle_2 \geq 0 \tag{2-33}$$

whenever $\phi, f_0\phi \in L^2(\Delta)$.

Proof.   We show that for any $\phi, \psi \in L^2(\Delta)$,

$$\lim_{\varepsilon \downarrow 0} \int_\Delta \int_\Delta \frac{\phi(s)\bar{\psi}(t)}{s-t+i\varepsilon}\, ds\, dt = \pi\langle (H_\Delta - iI)\phi,\psi\rangle_2. \tag{2-34}$$

To this end, set

$$\Phi(z) = \frac{1}{2\pi i} \int_\Delta \frac{\phi(t)}{t-z}\, dt, \qquad y \neq 0.$$

By Section 2.10, Theorem A,

$$\lim_{\varepsilon \downarrow 0} \Phi(x \pm i\varepsilon) = \tfrac{1}{2}[\pm\phi(x) - i(H_\Delta\phi)(x)] \tag{2-35}$$

a.e. on Δ and in the metric of $L^2(\Delta)$. Therefore,

$$\lim_{\varepsilon \downarrow 0} \int_\Delta \int_\Delta \frac{\phi(s)\bar{\psi}(t)}{s-t+i\varepsilon}\, ds\, dt = \int_\Delta \left(\lim_{\varepsilon \downarrow 0} \int_\Delta \frac{\phi(s)\, ds}{s-t+i\varepsilon} \right) \bar{\psi}(t)\, dt$$

$$= \langle \pi i(-\phi - iH_\Delta\phi),\psi\rangle_2,$$

and (2-34) follows.

If $\phi, f_0\phi \in L^2(\Delta)$, then by (2-34),

$$\lim_{\varepsilon \downarrow 0} \int_\Delta \int_\Delta \frac{f_0(s) - \bar{f}_0(t)}{s-t+i\varepsilon} \phi(s)\bar{\phi}(t)\, ds\, dt$$

$$= \pi\langle (H_\Delta - iI)f_0\phi,\phi\rangle_2 - \pi\langle (H_\Delta - iI)\phi,f_0\phi\rangle_2$$

$$= 2\pi\, \mathrm{Re}\langle (H_\Delta - iI)f_0\phi,\phi\rangle_2.$$

Thus (2-33) is equivalent to (2-30), and the result follows from 2.9.   ∎

2.12   Real Valued Functions and Loewner's Theorem

We prove the classical theorem of Loewner and a generalization to a union of intervals by specializing the generalized Loewner theorem 2.9, in its equivalent form in 2.11, to the case of real valued functions.

THEOREM A.   *Let $f_0(x)$ be a real valued measurable function on a Borel subset Δ of $(-\infty,\infty)$. There exists $f \in \mathscr{P}$ such that*

$$f(x) = f_0(x) \qquad a.e. \ on \ \Delta$$

if and only if

$$\lim_{\varepsilon \downarrow 0} \iint_{E(\varepsilon)} \frac{f_0(s) - f_0(t)}{s - t} \phi(s)\bar{\phi}(t)\,ds\,dt \geq 0 \qquad (2\text{-}36)$$

whenever $\phi, f_0\phi \in L^2(\Delta)$, where $E(\varepsilon) = \{(s,t): s,t \in \Delta \ and \ |s - t| > \varepsilon\}$ for each $\varepsilon > 0$.

Notice that the limit in (2-36) is of a different kind than that of (2-30). The proof shows that the limit in (2-36) exists as a real number whenever $\phi, f_0\phi \in L^2(\Delta)$, whether or not (2-36) holds.

Proof.   By 2.10, Theorem A, if $\phi, f_0\phi \in L^2(\Delta)$, then

$$\lim_{\varepsilon \downarrow 0} \iint_{E(\varepsilon)} \frac{f_0(s) - f_0(t)}{s - t} \phi(s)\bar{\phi}(t)\,ds\,dt$$

$$= \lim_{\varepsilon \downarrow 0} \int_\Delta \left(\int_\Delta \chi_{E(\varepsilon)}(s,t)\frac{f_0(s)\phi(s)}{s - t}\,ds \right)\bar{\phi}(t)\,dt$$

$$\quad - \lim_{\varepsilon \downarrow 0} \int_\Delta \left(\int_\Delta \chi_{E(\varepsilon)}(s,t)\frac{\phi(s)}{s - t}\,ds \right)f_0(t)\bar{\phi}(t)\,dt$$

$$= \pi\langle H_\Delta f_0\phi,\phi\rangle_2 - \pi\langle H_\Delta\phi,f_0\phi\rangle_2$$

$$= 2\pi\,\mathrm{Re}\langle H_\Delta f_0\phi,\phi\rangle_2$$

$$= 2\pi\,\mathrm{Re}\langle (H_\Delta - iI)f_0\phi,\phi\rangle_2.$$

Here $\chi_{E(\varepsilon)}$ is the characteristic function of $E(\varepsilon)$. The last equality holds because f_0 is real valued. Thus the result follows from the theorem in 2.11.   ∎

A holomorphic function f on Π is said to have an analytic continuation across an open subset Δ of $(-\infty, \infty)$ if $f = g|\Pi$, where g is holomorphic on an open set G containing $\Pi \cup \Delta$.

Suppose that $f \in \mathscr{P}$ and f has an analytic continuation across an open subset Δ of $(-\infty,\infty)$. Suppose further that this continuation is real valued on Δ. Then we may extend f to a holomorphic function on $\Pi \cup \tilde{\Pi} \cup \Delta$ such that

$$\bar{f}(\bar{z}) = f(z), \qquad z \in \Pi \cup \tilde{\Pi} \cup \Delta,$$

where $\tilde{\Pi} = \{z: \mathrm{Im}\,z < 0\}$. The next result characterizes this class of functions.

THEOREM B.   *Let f_0 be a real valued continuously differentiable function on an open subset Δ of $(-\infty,\infty)$. The following are equivalent:*

(i) *There exists $f \in \mathscr{P}$ such that f has an analytic continuation across Δ that coincides with f_0 on Δ.*

(ii) *For each continuous complex valued function ϕ on Δ that vanishes off a compact subset of Δ,*

$$\int_\Delta \int_\Delta K(s,t)\phi(s)\bar{\phi}(t)\,ds\,dt \geq 0, \qquad (2\text{-}37)$$

where for $s,t \in \Delta$,

$$K(s,t) = \begin{cases} [f_0(s) - f_0(t)]/(s - t), & s \neq t, \\ f'_0(t), & s = t. \end{cases} \qquad (2\text{-}38)$$

(iii) *For any finite sets $\{x_1,\ldots,x_n\} \subseteq \Delta$ and $\{c_1,\ldots,c_n\} \subseteq \mathbf{C}$,*

$$\sum_{j,k=1}^n K(x_j,x_k)c_j\bar{c}_k \geq 0, \qquad (2\text{-}39)$$

where $K(s,t)$ is defined as in (ii).

It is essential in the theorem that f_0 be real valued. The theorem is a generalization of Loewner's theorem (Löwner [1934]). Loewner's theorem is the special case of Theorem B in which Δ is an interval.

Proof.   By Theorem A, the following statements are equivalent:

(i') There exists $f \in \mathscr{P}$ whose boundary function satisfies $f(x) = f_0(x)$ a.e. on Δ.

(ii') The inequality (2-36) holds whenever $\phi, f_0\phi \in L^2(\Delta)$.

It is not hard to see that (i) is equivalent to (i'). The only point in question is that any function f as in (i') has an analytic continuation across Δ. This can be proved using a theorem of Carleman [1944], p. 40. A refined version of Carleman's theorem is given in Section 6.4.

We show that (ii) is equivalent to (ii'). A routine application of the mean value theorem shows that $K(s,t)$ is continuous on $\Delta \times \Delta$. Therefore in (2-36) we may replace "$\lim_{\varepsilon\downarrow 0} \iint_{E(\varepsilon)}$" by "$\iint_{\Delta\times\Delta}$". The equivalence of (ii) and (ii') now follows by a straightforward approximation argument.

To complete the proof, it remains to show that (ii) is equivalent to (iii). Assume (ii). It is sufficient to prove (iii) when $x_1,\ldots,x_n$ are distinct points of Δ, since this implies the general case of (iii). Let $c_1,\ldots,c_n \in \mathbf{C}$ be given. Set

$$\phi_\varepsilon(x) = \begin{cases} c_j/(2\varepsilon) & \text{if } |x - x_j| < \varepsilon, \\ 0 & \text{otherwise.} \end{cases}$$

For all sufficiently small $\varepsilon > 0$, ϕ_ε is well defined on Δ. By routine approximation, (2-37) holds for $\phi = \phi_\varepsilon$. We obtain (2-39) on letting $\varepsilon \downarrow 0$. Hence (ii) implies (iii). In the other direction, we deduce (ii) from (iii) by writing the integral in (2-37) as a limit of Riemann sums, each of which is nonnegative by (iii). ∎

2.13   Imaginary Functions and a Dual Result

The case of purely imaginary functions in the generalized Loewner theorem 2.9 or
the equivalent theorem of 2.11 is also interesting. A surprising result follows—
namely, a characterization of all nonnegative weight functions $g_0(x)$ on a real
Borel set Δ relative to which the operator H_Δ defined by (2-32) is bounded by 1.

THEOREM.   *Let g_0 be a nonnegative measurable function on a Borel subset Δ of
$(-\infty, \infty)$. The following are equivalent:*

(i) *There exists $f \in \mathscr{P}$ such that*

$$f(x) = ig_0(x) \qquad a.e. \ on \ \Delta.$$

(ii) *Whenever $\phi, g_0\phi \in L^2(\Delta)$,*

$$\int_\Delta |H_\Delta\phi|^2 g_0 \, dx \le \int_\Delta |\phi|^2 g_0 \, dx. \qquad (2\text{-}40)$$

Proof.   We first prove the theorem under the hypothesis that $g_0 \in L^1(\Delta) \cap
L^\infty(\Delta)$. We show that in this case, (i) and (ii) are equivalent to:

(iii) There exists $h \in \mathscr{P}$ such that

$$h(x) = -(H_\Delta g_0)(x) \qquad a.e. \ on \ \Delta. \qquad (2\text{-}41)$$

Note that $g_0 \in L^2(\Delta)$ because $L^1(\Delta) \cap L^\infty(\Delta) \subseteq L^2(\Delta)$, and therefore (2-41) is
meaningful. In the proof we use the following consequence of the convolution
theorem in 2.10: if $\phi \in L^2(\Delta) \cap L^\infty(\Delta)$, then

$$H_\Delta(\phi(H_\Delta\bar{\phi}) + (H_\Delta\phi)\bar{\phi}) = |H_\Delta\phi|^2 - |\phi|^2 \qquad a.e. \ on \ \Delta.$$

(ii) $\Leftrightarrow$ (iii) Assume (ii), and set $f_0 = -H_\Delta g_0$. For any $\phi \in L^2(\Delta) \cap L^\infty(\Delta)$, we
have $\phi, g_0\phi \in L^2(\Delta)$, so

$$\begin{aligned}
\operatorname{Re}\langle(H_\Delta - iI)f_0\phi,\phi\rangle_2 &= \tfrac{1}{2}\langle H_\Delta(f_0\phi),\phi\rangle_2 + \tfrac{1}{2}\langle\phi,H_\Delta(f_0\phi)\rangle_2 \\
&= -\tfrac{1}{2}\langle f_0\phi,H_\Delta\phi\rangle_2 - \tfrac{1}{2}\langle H_\Delta\phi,f_0\phi\rangle_2 \\
&= \tfrac{1}{2}\int_\Delta (\phi(H_\Delta\bar{\phi}) + (H_\Delta\phi)\bar{\phi})H_\Delta g_0 \, dx \\
&= -\tfrac{1}{2}\int_\Delta \{H_\Delta(\phi(H_\Delta\bar{\phi}) + (H_\Delta\phi)\bar{\phi})\}g_0 \, dx \\
&= -\tfrac{1}{2}\int_\Delta (|H_\Delta\phi|^2 - |\phi|^2)g_0 \, dx \\
&\ge 0.
\end{aligned}$$

By approximation, (2-33) holds for arbitrary $\phi \in L^2(\Delta)$ such that $f_0\phi \in L^2(\Delta)$. By
2.11, (iii) holds.

Reversing the above argument, we see that if (iii) holds, then (2-40) holds at least for all $\phi \in L^2(\Delta) \cap L^\infty(\Delta)$. By approximation, (2-40) holds for all $\phi \in L^2(\Delta)$ such that $g_0 \phi \in L^2(\Delta)$. Hence (ii) and (iii) are equivalent.

(i) $\Leftrightarrow$ (iii) Assume (iii), and choose $h \in \mathscr{P}$ such that (2-41) holds. Then

$$f(z) = \frac{1}{\pi} \int_\Delta \frac{g_0(t)}{t - z} \, dt + h(z), \qquad z \in \Pi, \tag{2-42}$$

defines a function in $\mathscr{P}$. By Theorem A (Section 2.10),

$$f(x) = \tfrac{1}{2} \cdot 2i\{g_0(x) - i(H_\Delta g_0)(x)\} + h(x)$$

$$= ig_0(x) \qquad \text{a.e. on } \Delta,$$

so (i) holds.

Conversely, let (i) hold and choose $f \in \mathscr{P}$ such that $f(x) = ig_0(x)$ a.e. on Δ. Define h on Π so that (2-42) holds. Since g_0 is essentially bounded on Δ, Im h is bounded below on Π and $\exp(ih) \in H^\infty(\Pi)$. But

$$\text{Im } h(z) = \text{Im } f(z) - \frac{y}{\pi} \int_\Delta \frac{g_0(t)}{(t - x)^2 + y^2} \, dt, \qquad y > 0,$$

where

$$\lim_{y \downarrow 0} \frac{y}{\pi} \int_\Delta \frac{g_0(t)}{(t - x)^2 + y^2} \, dt = \begin{cases} g_0(x) & \text{a.e. on } \Delta, \\ 0 & \text{a.e. on } (-\infty, \infty) \backslash \Delta. \end{cases}$$

It follows that

$$\text{Im } h(x) = \begin{cases} 0 & \text{a.e. on } \Delta, \\ \text{Im } f(x) & \text{a.e. on } (-\infty, \infty) \backslash \Delta, \end{cases}$$

and thus Im $h(x) \geq 0$ a.e. on $(-\infty, \infty)$. Therefore the boundary function of $\exp(ih)$ is bounded by 1 a.e. on $(-\infty, \infty)$, and so $|\exp(ih)| \leq 1$ on Π. Hence $h \in \mathscr{P}$. Since

$$h(x) = f(x) - \tfrac{1}{2} \cdot 2i\{g_0(x) - i(H_\Delta g_0)(x)\}$$

$$= -(H_\Delta g_0)(x)$$

a.e. on Δ, (iii) follows.

We have proved the theorem in the special case in which $g_0 \in L^1(\Delta) \cap L^\infty(\Delta)$. Now consider the general case. Choose Borel subsets $\Delta_1, \Delta_2, \ldots$ of Δ such that $\Delta_1 \subseteq \Delta_2 \subseteq \ldots$, $\bigcup_1^\infty \Delta_n = \Delta$, and the restriction $g_0^{(n)}$ of g_0 to Δ_n belongs to $L^1(\Delta_n) \cap L^\infty(\Delta_n)$ for each n.

Assume (i). By the special case of the theorem proved above, for each $n = 1, 2, 3, \ldots$,

$$\int_{\Delta_n} |H_{\Delta_n} \phi|^2 g_0^{(n)} \, dx \leq \int_{\Delta_n} |\phi|^2 g_0^{(n)} \, dx$$

whenever ϕ, $g_0^{(n)}\phi \in L^2(\Delta_n)$. A straightforward approximation argument shows that (ii) follows.

Conversely, let (ii) hold. By the special case of the theorem proved above, for each $n = 1,2,3,\ldots$, there exists $f_n \in \mathscr{P}$ such that $f_n(x) = ig_0^{(n)}(x)$ a.e. on Δ_n. If Δ is a Lebesgue null set, the theorem is trivial. Otherwise we may assume that Δ_1 is not a Lebesgue null set. Since functions in $\mathscr{P}$ are determined by their boundary functions on sets of positive measure, $f_1 = f_2 = \cdots$ on Π. Denoting this function f, we obtain $f \in \mathscr{P}$ and $f(x) = ig_0(x)$ a.e. on Δ. The proof is now complete. $\blacksquare$

Examples and Addenda

1. *Monotone matrix and operator functions and a theorem of Loewner.* Let f be a real valued Borel function on a Borel set $\Delta \subseteq (-\infty,\infty)$. Assume that f is bounded on all compact subsets of Δ. We call f a *monotone operator function* if whenever A,B are bounded self-adjoint operators on a Hilbert space $\mathscr{H}$ with spectra $\mathrm{sp}(A) \subseteq \Delta$, $\mathrm{sp}(B) \subseteq \Delta$, and $A \leq B$, then $f(A) \leq f(B)$; f is a *monotone matrix function* if this condition holds with the added restriction that $\dim \mathscr{H} < \infty$. We use the standard functional calculus for self-adjoint operators: if A has spectral representation $A = \int_{\mathrm{sp}(A)} t\, dE(t)$, then $f(A) = \int_{\mathrm{sp}(A)} f(t)\, dE(t)$. A monotone matrix or operator function is monotone in the usual sense, but the converse of this statement is false.

LOEWNER'S THEOREM (*Löwner* [1934]). *Let f be a real Borel function on an open interval $\Delta \subseteq (-\infty,\infty)$ that is bounded on all compact subsets of Δ.*

   I. *If f is a monotone matrix function on Δ, then $f = g|\Delta$, where g is holomorphic on $\Pi \cup \tilde{\Pi} \cup \Delta$ and $\mathrm{Im}\, g \geq 0$ on Π.*

  II. *If $f = g|\Delta$, where g is holomorphic on $\Pi \cup \tilde{\Pi} \cup \Delta$ and $\mathrm{Im}\, g \geq 0$ on Π, then f is a monotone operator function on Δ.*

Here $\Pi,\tilde{\Pi}$ are the open upper and lower half-planes, respectively. A corollary of the theorem is that every monotone matrix function on an open interval Δ is a monotone operator function on Δ.

Examples. (i) $x^{1/2}$ and $\log x$ are monotone operator functions on $(0,\infty)$.
(ii) x^2 and e^x are not monotone operator functions on any interval (a,b).

We outline a proof of Loewner's theorem in nos. 2–4 below.

2. *Proof of Loewner's theorem, part* II.   The proof of part II of the theorem is carried out in steps as follows.

  (i) If A,B are bounded self-adjoint operators on a Hilbert space $\mathscr{H}$ such that $B \geq A \geq \delta I$ for some $\delta > 0$, then $B^{-1} \leq A^{-1}$.
  (ii) If c is a real number not in the open interval $\Delta \subseteq (-\infty,\infty)$, then $f_c(x) = 1/(c - x)$ is a monotone operator function on Δ.
 (iii) Let $f = g|\Delta$ as in part II of Loewner's theorem. Consider the Nevanlinna

representation of $g \mid \Pi$ (Appendix, Section 6):

$$g(z) = b + cz + \frac{1}{\pi} \int\limits_{-\infty}^{\infty} \left(\frac{1}{t - z} - \frac{t}{1 + t^2} \right) d\mu(t), \qquad z \in \Pi.$$

Here $b = \bar{b}, c \geq 0$, and $\int_{-\infty}^{\infty} (1 + t^2)^{-1} d\mu(t) < \infty$. By the Stieltjes inversion formula (Appendix, Section 6), $\mu(\Delta) = 0$. It follows that

$$f(x) = b + cx + \frac{1}{\pi} \int\limits_{(-\infty,\infty)\backslash\Delta} \left(\frac{1}{t - x} - \frac{t}{1 + t^2} \right) d\mu(t), \qquad x \in \Delta.$$

Using this representation of f and (ii), we see that f is a monotone operator function on Δ.

3. *Some lemmas.* We prepare the way for the proof of part I of Loewner's theorem with some preliminary results.

LEMMA A.   *Let $A = \sum_1^m \lambda_j E_j$, $B = \sum_1^n \mu_k F_k$, where $\lambda_1, \ldots, \lambda_m$, $\mu_1, \ldots, \mu_n$ are real numbers, and $E_1, \ldots, E_m$, $F_1, \ldots, F_n$ are projections on a Hilbert space $\mathcal{H}$ such that $E_j E_k = 0$ for $j \neq k$ $(j,k = 1, \ldots, m)$, $F_j F_k = 0$ for $j \neq k$ $(j,k = 1, \ldots, n)$, and $\sum_1^m E_j = \sum_1^n F_k = I$. Then for any Borel function f whose domain includes all of the λ's and μ's,*

$$f(B) - f(A) = \sum_{j=1}^m \sum_{k=1}^n {}' \frac{f(\lambda_j) - f(\mu_k)}{\lambda_j - \mu_k} E_j(B - A)F_k, \qquad (2\text{-}43)$$

where the ' indicates that when $\lambda_j - \mu_k = 0$, the difference quotient $[f(\lambda_j) - f(\mu_k)]/(\lambda_j - \mu_k)$ may be replaced by any number whatever, the choice being irrelevant because $E_j(B - A)F_k = 0$.

Proof.   In the sum on the right of (2-43), substitute $E_j A = \lambda_j E_j$ and $BF_k = \mu_k F_k$, and then simplify.   ∎

LEMMA B.   *Let A,B be self-adjoint operators on a Hilbert space of finite dimension n. Assume that the eigenvalues of A are simple. Let $A = \sum_1^n \lambda_j P_j$ be the spectral representation of A. Then for all sufficiently small real ε, the eigenvalues of $A^{(\varepsilon)} = A + \varepsilon B$ are all simple, and the spectral representation of $A^{(\varepsilon)}$ can be written in such a form $A^{(\varepsilon)} = \sum_1^n \lambda_j^{(\varepsilon)} P_j^{(\varepsilon)}$ that for each $j = 1, \ldots, n$, $\lambda_j^{(\varepsilon)} \to \lambda_j$ and $P_j^{(\varepsilon)} \to P_j$ as $\varepsilon \to 0$.*

Proof.   Since the eigenvalues of A are simple, $\lambda_1, \ldots, \lambda_n$ are distinct numbers, and $P_1, \ldots, P_n$ are rank one projections. Choose circles $\gamma_1, \ldots, \gamma_n$ about $\lambda_1, \ldots, \lambda_n$ small enough that their interiors are disjoint. Then for each $j = 1, \ldots, n$,

$$P_j = \frac{1}{2\pi i} \int\limits_{\gamma_j} (\zeta I - A)^{-1} d\zeta, \qquad \lambda_j = \text{tr } P_j A,$$

where tr indicates trace. For all sufficiently small real ε, $A^{(\varepsilon)}$ has spectral representation $A^{(\varepsilon)} = \sum_1^n \lambda_j^{(\varepsilon)} P_j^{(\varepsilon)}$, where

$$P_j^{(\varepsilon)} = \frac{1}{2\pi i} \int_{\gamma_j} (\zeta I - A^{(\varepsilon)})^{-1}\, d\zeta,$$

$$\lambda_j^{(\varepsilon)} = \text{tr}\, P_j^{(\varepsilon)} A^{(\varepsilon)},$$

$j = 1,\ldots,n$; each $P_j^{(\varepsilon)}$ is a rank one projection, and the associated eigenvalue $\lambda_j^{(\varepsilon)}$ lies in the interior of γ_j. Clearly, $\lambda_j^{(\varepsilon)} \to \lambda_j$ and $P_j^{(\varepsilon)} \to P_j$ as $\varepsilon \to 0$. ∎

LEMMA C. *Let f be a locally bounded Borel function on $\Delta = (a,b)$. Let $\phi \geq 0$ be a C^∞ function with compact support in $(-1,1)$ such that $\int_{-1}^1 \phi\, dt = 1$. For each ε, $0 < \varepsilon < (b - a)/2$, set*

$$f_\varepsilon(x) = \int_{-1}^1 \phi(t) f(x + \varepsilon t)\, dt$$

on $\Delta_\varepsilon = (a + \varepsilon, b - \varepsilon)$. Then $f_\varepsilon \in C^\infty(\Delta_\varepsilon)$, and $f_\varepsilon(x) \to f(x)$ at every point $x \in \Delta$ where f is continuous.

Proof. Extend ϕ, f to $(-\infty, \infty)$ by setting both equal to zero off their domains. Then

$$f_\varepsilon(x) = \int_{-\infty}^\infty \varepsilon^{-1} \phi((s - x)/\varepsilon) f(s)\, ds, \qquad x \in \Delta_\varepsilon.$$

With this representation the proof becomes a pleasant (and familiar) exercise in real analysis. ∎

4. *Proof of Loewner's theorem, part* I. Let f be a monotone matrix function on Δ. As a first case, suppose that f is continuously differentiable on Δ. Then

$$K(x,y) = \begin{cases} [f(x) - f(y)]/(x - y), & x \neq y, \\ f'(x), & x = y, \end{cases}$$

is continuous on $\Delta \times \Delta$. We show that

$$\sum_{j,k=1}^m K(\lambda_j, \lambda_k) c_j \bar{c}_k \geq 0 \tag{2-44}$$

whenever $\{\lambda_1,\ldots,\lambda_n\} \subseteq \Delta$ and $\{c_1,\ldots,c_n\} \subseteq \mathbf{C}$. Without loss of generality we can assume that $\lambda_1,\ldots,\lambda_n$ are distinct.

Let $\mathscr{H}$ be a Hilbert space with dim $\mathscr{H} = n$, and let $e_1,\ldots,e_n$ be an orthonormal basis for $\mathscr{H}$. Let $A = \sum_1^n \lambda_j P_j$, where $P_j = \langle \cdot, e_j \rangle e_j$ is the projection of $\mathscr{H}$ on the span of e_j, $j = 1,\ldots,n$. Then A is a self-adjoint operator with $\text{sp}(A) \subseteq \Delta$. Set $B = A + \varepsilon Q$, where $Q = \langle \cdot, v \rangle v$, $v = e_1 + \cdots + e_n$, and $\varepsilon > 0$. For all sufficiently small ε, $\text{sp}(B) \subseteq \Delta$. Since $A \leq B$, $f(A) \leq f(B)$. By Lemma B in no. 3

above, the spectral representation of B can be written in such a form $B = \sum_1^n \lambda_j^{(\varepsilon)} P_j^{(\varepsilon)}$ that $\lambda_j^{(\varepsilon)} \to \lambda_j$ and $P_j^{(\varepsilon)} \to P_j$ as $\varepsilon \downarrow 0$. By Lemma A in no. 3,

$$0 \le f(B) - f(A) = \sum_{j,k=1}^m K(\lambda_j, \lambda_k^{(\varepsilon)}) P_j \varepsilon Q P_k^{(\varepsilon)}.$$

Dividing by ε, and then letting $\varepsilon \downarrow 0$, we obtain

$$\sum_{j,k=1}^m K(\lambda_j, \lambda_k) P_j Q P_k \ge 0.$$

Put $u = \sum_1^n \bar{c}_j e_j$. Then the inequality

$$\sum_{j,k=1}^m K(\lambda_j, \lambda_k) \langle P_j Q P_k u, u \rangle \ge 0$$

simplifies to (2-44).

By Theorem B (Section 2.12), $f = g|\Delta$, where g is holomorphic on an open set containing $\Pi \cup \Delta$ and $\operatorname{Im} g \ge 0$ on Π. Since $f = g|\Delta$ is real valued, by reflection g is holomorphic on $\Pi \cup \tilde{\Pi} \cup \Delta$. This completes the proof of Loewner's theorem, part I, in the special case where f is assumed to be continuously differentiable on Δ.

The general case of Loewner's theorem, part I, then follows with the aid of Lemma C in no. 3 and a compactness argument. The details are straightforward, and we omit them.   ∎

The preceding proof of Loewner's theorem is closely related to a method of Yu. L. Daleckiĭ and S. G. Kreĭn, which yields a Taylor expansion for functions with an operator argument. Daleckiĭ [1957] gives an account of the method; he remarks that connections with Loewner's theorem were pointed out by M. G. Kreĭn. The method is also used by Davis [1963].

5. *Monotone operator functions on more general sets.* It is natural to ask if Loewner's theorem in no. 1 above generalizes to sets other than open intervals. The following results, which we state without proof, show that, in some sense, the natural domain of a monotone operator function is indeed an interval.

THEOREM (*Šmul'jan* [1965]). *Let f be a monotone operator function on $\{a\} \cup (b,c)$, where $-\infty < a < b < c \le \infty$. Then there is a monotone operator function g on (a,c) such that $g(x) = f(x)$ for all $x \in (b,c)$ and $g(a + 0) \ge f(a)$.*

THEOREM (*Chandler* [1976]). *Let f be a monotone operator function on an open set $\Delta \subseteq (-\infty, \infty)$. Then $f = g|\Delta$, where g is a monotone operator function on the convex hull of Δ.*

For additional literature, see Bendat and Sherman [1955]; Davis [1957], [1963]; Donoghue [1974], [1980]; Hansen and Pedersen [1981/82]; Korányi [1983]; and Sparr [1980].

Notes

2.2   Nudel'man [1977].

2.3   Rosenblum and Rovnyak [1980]. The result may be viewed as a fusion of methods of Nudel'man [1977] and Sarason [1967]. Theorem 2.3 has a generalization to operator valued functions, with similar applications. See Rosenblum and Rovnyak [1982].

2.4–2.5   Carathéodory [1907], [1911] initiated the study of coefficient problems. Toeplitz [1911b] showed how Carathéodory's theorem on the coefficients of a holomorphic function with positive real part in the unit disk can be stated in terms of the nonnegativity of quadratic forms. The classical Carathéodory-Fejér theorem appears in Carathéodory and Fejér [1911].

Pick first considered his interpolation problem as a generalization of the coefficient problems studied by Carathéodory and others. Pick's theorem as stated in 2.4 appears in Pick [1916] in an equivalent though easily recognizable form. He wrote two other papers on the subject: Pick [1918], [1920]. Due to the turmoil of the world war, Pick's work was not known to Nevanlinna when he independently investigated the same problem and wrote the first of several massive memoirs on the subject: Nevanlinna [1919]. In Nevanlinna's view, the problem was a natural development not only of theorems on coefficient problems, but also of Schwarz's lemma and its generalizations. Nevanlinna's criteria for the existence of a solution are of a different form than Pick's. Nevanlinna showed moreover how to construct solutions. We mention also Schur [1917], [1918], who earlier used a similar method. In addition, see Nevanlinna [1922a,b], [1929]. An appreciation of Nevanlinna's work on the interpolation problem has been written by Pfluger [1982]. Nevanlinna [1919] contains a section of more than twenty pages on the historical evolution of the problem. The form of the Pick-Nevanlinna theorem stated in 2.4 is due to Kreĭn and Rehtman [1938].

The ideas of Pick and Nevanlinna have inspired many subsequent writers, and there exist numerous accounts in the literature from both classical and modern perspectives. See, for example, Ahiezer [1961], Garnett [1981], Kreĭn and Nudel'man [1973], Sarason [1967], and Walsh [1935], as well as the sources for Loewner's theory cited below. Heins [1984] has compiled a working bibliography on the Pick-Nevanlinna problem that covers ground well beyond the scope of our discussion. See also Stray [1984]. Algorithms suitable for numerical calculation are discussed in Allison and Young [1983] and Young [1984].

2.6   A simultaneous generalization of the Pick-Nevanlinna and Carathéodory-Fejér theorems was first proved by Helton [1978]. Helton also uses the lifting theorem, in the original form due to Sarason [1967]. Our version appeared in Rosenblum and Rovnyak [1980].

2.7   Rosenblum and Rovnyak [1975I], [1980]. The paper [1975I] contains some additional theorems of a similar nature.

2.9   This result was first obtained in Rosenblum and Rovnyak [1975I] by a Toeplitz operator method. Independently, FitzGerald [1978] discovered a special case using function theoretic methods. A different generalization of Loewner's theorem is given in FitzGerald [1984].

2.12   Loewner published his theory in a penetrating paper, Löwner [1934], and he wrote another paper, Loewner [1950], on related questions. Loewner was evidently led to his ideas, in part, through speculations on relativity theory (verbal communication by Loewner's student, Carl FitzGerald; Bendat and Sherman [1955], pp. 58–59). A connection with quantum mechanics is shown in Wigner and von Neumann [1954]. Bendat and Sherman [1955] cite references to other physical applications and also show a connection with the Hamburger moment problem. An elegant approach to the Pick-Nevanlinna and Loewner theorems via operator theory and reproducing kernel Hilbert spaces was developed by Korányi [1955], [1956], Sz.-Nagy [1956], and Sz.-Nagy and Korányi [1956], [1958]. For modern accounts of the Pick-Nevanlinna and Loewner theories, see Donoghue [1974], FitzGerald [1977], and FitzGerald and Horn [1982].

The generalization of the classical Loewner theorem to a union of intervals is a special case of a theorem in Rosenblum and Rovnyak [1975I], pp. 118–119. A proof based on reproducing kernel Hilbert spaces has recently been found by Donoghue [1983].

2.13   Rosenblum and Rovnyak [1975III]. A related result is given in Rosenblum and Rovnyak [1974]. Hunt, Muckenhoupt, and Wheeden [1973] have determined all nonnegative weight functions on the real line relative to which the Hilbert transform is a bounded operator in the p-norm.

Additional Literature

GENERALIZATIONS TO OPERATOR VALUED FUNCTIONS.   Delsarte, Genin, and Kamp [1979a,b], Erwe [1966], Fedčina [1972], [1975], Korányi [1956], Rosenblum and Rovynak [1982], Sz.-Nagy and Korányi [1958].

SIGNED INTERPOLATION AND INTERPOLATION BY MEROMORPHIC FUNCTIONS.   Ball [1983], Ball and Helton [1982b], [1983], Clark [1968], Kreĭn and Langer [1977], Nudel'man [1981].

THEORY OF ADAMJAN, AROV AND KREĬN; THEORY OF KOVALISHINA AND POTAPOV.   These theories provide penetrating studies and far reaching generalizations of the Carathéodory-Fejér and Pick-Nevanlinna theorems. Concerning the former, see Adamjan, Arov and Kreĭn [1971a,b], Garnett [1981] (Chapter IV), and Stray [1981]. For the latter, see Kovalishina [1983] and the monograph of Kovalishina and Potapov [1982].

GENERALIZATIONS TO FUNCTIONS OF SEVERAL VARIABLES.   Beatrous [1981], Dautov and Khudaĭberganov [1982], Hamilton [1983], Korányi [1961], Korányi and Pukánszky [1963], Vasudeva [1973].

ENGINEERING AND PHYSICAL APPLICATIONS.   Delsarte, Genin, and Kamp [1981], Saal [1984].

Chapter 3

Factorization of Toeplitz Operators

3.1 Introduction

Throughout this chapter we assume that S is a shift operator on a Hilbert space $\mathcal{H}$. We write $P_0 = I - SS^*$ for the projection of $\mathcal{H}$ on $\mathcal{K} = \ker S^*$. Analytic, inner, outer, and S-constant operators are defined relative to S as in Section 1.6.

DEFINITION.   *An operator $T \in \mathcal{B}(\mathcal{H})$ is* Toeplitz, *or, more precisely,* S-Toeplitz, *if* $S^*TS = T$.

Example 1.   Let

$$S_1 : (c_0, c_1, c_2, \ldots) \to (0, c_0, c_1, \ldots)$$

on l^2, and let T be a bounded linear operator on l^2 with matrix $[w_{jk}]_{j,k=0}^{\infty}$. Thus $T : (a_0, a_1, a_2, \ldots) \to (b_0, b_1, b_2, \ldots)$ if and only if

$$b_j = \sum_{k=0}^{\infty} w_{jk} a_k, \qquad j = 0, 1, 2, \ldots.$$

The matrix of S^*TS is $[w_{j+1,k+1}]_{j,k=0}^{\infty}$. Hence T is an S_1-Toeplitz operator if and only if its matrix has the form $[c_{j-k}]_{j,k=0}^{\infty}$ for some sequence $\{c_n\}_{-\infty}^{\infty}$. Such a matrix is called a *Toeplitz matrix*.

Example 2.   Let S_2 be multiplication by $e^{i\theta}$ on $H^2(\Gamma)$. Each $w \in L^{\infty}(\sigma)$ induces an S_2-Toeplitz operator $T(w)$ on $H^2(\Gamma)$ defined by

$$T(w)f = Pwf, \qquad f \in H^2(\Gamma),$$

where P is the projection of $L^2(\sigma)$ on $H^2(\Gamma)$. Every Toeplitz operator on $H^2(\Gamma)$ has this form (Examples and Addenda, no. 1).

Example 3.   In general, examples of Toeplitz operators are easily constructed from analytic operators. If $A, C \in \mathcal{B}(\mathcal{H})$ are analytic, then the operators A, C^*, C^*A are Toeplitz. For example, if $T = C^*A$, then $S^*TS = S^*C^*AS = C^*S^*SA = C^*A = T$.

The main ideas in Chapter 3 can be illustrated in the situation of Example 2. If $w = a \in H^{\infty}(\Gamma)$, then $T(a)$ is the analytic operator multiplication by a on $H^2(\Gamma)$. If $a, c \in H^{\infty}(\Gamma)$, then

$$T(c)\, T(a) = T(ca) \qquad \text{and} \qquad T(c)^*T(a) = T(\bar{c}a).$$

Therefore factorization properties of functions on the circle may be interpreted as analogous properties of Toeplitz operators. Our program is to reverse this procedure. We shall develop an abstract factorization theory for Toeplitz operators in this chapter. Then in Chapters 5 and 6 we will interpret the abstract theory in concrete situations, thus obtaining factorization theorems for operator valued functions.

In Section 3.2 we associate with any $T \in \mathcal{B}(\mathcal{H})$ a matrix $[A_{jk}]_{j,k=0}^{\infty}$ of operators in $\mathcal{B}(\mathcal{K})$, and we show that T is a Toeplitz operator if and only if the matrix has the Toeplitz form. The central result in Chapter 3 is Theorem 3.4, which solves the abstract Szegö problem stated in Section 1.8. We give an inner-outer factorization theorem for analytic operators in Section 3.6. In Sections 3.10 and 3.11 we characterize outer operators in terms of extremal properties. Additional examples of Toeplitz operators and the concrete spectral theory of self-adjoint Toeplitz operators are outlined in the Examples and Addenda at the end of the chapter.

3.2   Operator Matrices

To each $T \in \mathcal{B}(\mathcal{H})$ we associate a matrix of operators in $\mathcal{B}(\mathcal{K})$: $T \sim [A_{jk}]_{j,k=0}^{\infty}$, where

$$A_{jk} = P_0 S^{*j} T S^k P_0 | \mathcal{K}, \qquad j,k \geq 0. \tag{3-1}$$

THEOREM A.   *Let* $T \in \mathcal{B}(\mathcal{H})$ *and* $T \sim [A_{jk}]_{j,k=0}^{\infty}$. *For any* $f \in \mathcal{H}$, *if* $f = \sum_0^{\infty} S^j a_j$ *as in* (1-4), *then* $Tf = \sum_0^{\infty} S^j b_j$, *where*

$$\begin{bmatrix} A_{00} & A_{01} & A_{02} & \cdots \\ A_{10} & A_{11} & A_{12} & \cdots \\ A_{20} & A_{21} & A_{22} & \cdots \\ & & \cdots & \end{bmatrix} \begin{bmatrix} a_0 \\ a_1 \\ a_2 \\ \vdots \end{bmatrix} = \begin{bmatrix} b_0 \\ b_1 \\ b_2 \\ \vdots \end{bmatrix}. \tag{3-2}$$

Proof.   If $Tf = \sum_0^{\infty} S^j b_j$ as in (1-4), then by (1-5),

$$b_j = P_0 S^{*j} T f = P_0 S^{*j} T \sum_{k=0}^{\infty} S^k a_k = \sum_{k=0}^{\infty} A_{jk} a_k, \qquad j \geq 0.$$

The sums are strongly convergent, and the sequence of relations is equivalent to (3-2). ∎

The correspondence $T \sim [A_{jk}]_{j,k=0}^{\infty}$ is clearly linear and well behaved with respect to adjoints: if $T \sim [A_{jk}]_{j,k=0}^{\infty}$, then $T^* \sim [B_{jk}]_{j,k=0}^{\infty}$, where $B_{jk} = A_{kj}^*$ for all $j,k \geq 0$.

THEOREM B.   *If* $T_1 \sim [A_{jk}]_{j,k=0}^{\infty}$ *and* $T_2 \sim [B_{jk}]_{j,k=0}^{\infty}$, *then* $T_1 T_2 \sim [C_{jk}]_{j,k=0}^{\infty}$, *where*

$$C_{jk} = \sum_{l=0}^{\infty} A_{jl} B_{lk}, \qquad j,k \geq 0, \tag{3-3}$$

with convergence of the series in the strong operator topology.

Proof.   By the Wold decomposition 1.3 (operator version),

$$I = \sum_{l=0}^{\infty} S^l P_0 S^{*l}$$

with convergence in the strong operator topology. Hence for any $j,k \geq 0$,

$$
\begin{aligned}
C_{jk} &= P_0 S^{*j} T_1 T_2 S^k P_0 | \mathcal{K} \\
&= P_0 S^{*j} T_1 \sum_{l=0}^{\infty} S^l P_0 S^{*l} T_2 S^k P_0 | \mathcal{K} \\
&= \sum_{l=0}^{\infty} A_{jl} B_{lk}
\end{aligned}
$$

with convergence in the strong operator topology. ∎

THEOREM C. *An operator $T \in \mathcal{B}(\mathcal{H})$ is Toeplitz if and only if its matrix has the form*

$$
[A_{j-k}]_{j,k=0}^{\infty} = \begin{bmatrix} A_0 & A_{-1} & A_{-2} & \cdots \\ A_1 & A_0 & A_{-1} & \cdots \\ A_2 & A_1 & A_0 & \cdots \\ & & \cdots & \end{bmatrix}.
\tag{3-4}
$$

In this case,

$$
A_j = \begin{cases} P_0 S^{*j} T P_0 | \mathcal{K}, & j \geq 0, \\ P_0 T S^{|j|} P_0 | \mathcal{K}, & j < 0. \end{cases}
\tag{3-5}
$$

A matrix of the form (3-4) is called a *Toeplitz matrix*.

Proof. Let T be Toeplitz with matrix $[A_{jk}]_{j,k=0}^{\infty}$. Then $S^{*k} T S^k = T$ for all $k \geq 0$. Hence

$$
A_{jk} = P_0 S^{*j} T S^k P_0 = \begin{cases} P_0 S^{*j-k} T P_0 | \mathcal{K} & \text{if } j \geq k, \\ P_0 T S^{k-j} P_0 | \mathcal{K} & \text{if } j < k. \end{cases}
$$

Thus T has the matrix (3-4) where the entries are defined by (3-5).

Conversely, let the matrix of T have the form (3-4). Then by (3-1) the operators T and $S^* TS$ have the same matrix. Hence $T = S^* TS$ and T is Toeplitz. ∎

COROLLARY. *An operator $A \in \mathcal{B}(\mathcal{H})$ is analytic if and only if its matrix has the form*

$$
\begin{bmatrix} A_0 & 0 & 0 & \cdots \\ A_1 & A_0 & 0 & \cdots \\ A_2 & A_1 & A_0 & \cdots \\ & & \cdots & \end{bmatrix}.
\tag{3-6}
$$

In this case

$$
A_j = P_0 S^{*j} A P_0 | \mathcal{K}, \qquad j \geq 0.
\tag{3-7}
$$

Moreover, A is S-constant if and only if $A_j = 0$ for all $j \geq 1$, that is, the matrix of A has the form

$$
\text{diag}\{A_0, A_0, A_0, \ldots\}.
\tag{3-8}
$$

3.3   Lowdenslager's Isometry

Let $T \in \mathcal{B}(\mathcal{H})$ be a nonnegative Toeplitz operator. Since $T \geq 0$, there exists a unique nonnegative square root $T^{1/2}$. Since T is Toeplitz, $S^*TS = T$ and so for any $f \in \mathcal{H}$,

$$\|T^{1/2}Sf\|^2 = \langle S^*TSf, f \rangle = \langle Tf, f \rangle = \|T^{1/2}f\|^2.$$

It follows that there exists a unique isometry on $T^{1/2}\mathcal{H}$ to $T^{1/2}\mathcal{H}$ that maps $T^{1/2}f$ to $T^{1/2}Sf$ for each $f \in \mathcal{H}$. The extension by continuity of this isometry to $\overline{T^{1/2}\mathcal{H}}$ plays a central role in what follows.

DEFINITION.   Let $T \in \mathcal{B}(\mathcal{H})$ be a nonnegative Toeplitz operator. We set

$$\mathcal{H}_T = \overline{T^{1/2}\mathcal{H}}$$

and view $\mathcal{H}_T$ as a Hilbert space in the inner product of $\mathcal{H}$. Lowdenslager's isometry is the unique isometry S_T on $\mathcal{H}_T$ such that

$$S_T(T^{1/2}f) = T^{1/2}Sf \text{ for all } f \in \mathcal{H}.$$

We use Lowdenslager's isometry to give the following solution to the abstract Szegö problem of Section 1.8.

3.4   Theorem. Solution of the Abstract Szegö Problem

Let $T \in \mathcal{B}(\mathcal{H})$ be a nonnegative Toeplitz operator. The following assertions are equivalent:

   (i) $T = A^*A$ for some analytic operator $A \in \mathcal{B}(\mathcal{H})$;
   (ii) Lowdenslager's isometry S_T is a shift operator;
   (iii) for all vectors c in some dense subset $\mathcal{D}$ of $\mathcal{K}$,

$$\lim_{n \to \infty} \left(\sup\{|\langle Tc, S^n f \rangle| : f \in \mathcal{H}, \langle Tf, f \rangle = 1\} \right) = 0. \qquad (3\text{-}9)$$

In this case there exists an outer operator $A \in \mathcal{B}(\mathcal{H})$ such that $T = A^*A$ and in the matrix (3-6) for A, $A_0 \geq 0$.

Proof.   (i) $\Rightarrow$ (iii) Let $T = A^*A$, where $A \in \mathcal{B}(\mathcal{H})$ is analytic. If $c \in \mathcal{K}$, $f \in \mathcal{H}$, and $\langle Tf, f \rangle = 1$, then

$$|\langle Tc, S^n f \rangle|^2 = |\langle A^*Ac, S^n f \rangle|^2 = |\langle S^{*n}Ac, Af \rangle|^2$$

$$\leq \|S^{*n}Ac\|^2 \|Af\|^2 = \|S^{*n}Ac\|^2 \langle Tf, f \rangle = \|S^{*n}Ac\|^2$$

for each $n = 0, 1, 2, \ldots$. Since S is a shift operator, $\|S^{*n}Ac\| \to 0$, so (iii) holds with $\mathcal{D} = \mathcal{K}$.

   (iii) $\Rightarrow$ (ii) Assume (iii). Claim: for each $c \in \mathcal{D}$ and $n \geq 0$,

$$\|S_T^{*n}T^{1/2}c\| = \sup\{|\langle Tc, S^n f \rangle| : f \in \mathcal{H}, \langle Tf, f \rangle = 1\}.$$

To see this, note that for any $f \in \mathcal{H}$,

$$\langle Tc, S^n f \rangle = \langle T^{1/2}c, T^{1/2}S^n f \rangle = \langle T^{1/2}c, S_T^n T^{1/2} f \rangle$$
$$= \langle S_T^{*n}T^{1/2}c, T^{1/2}f \rangle.$$

The claim then follows from the fact that the set of vectors $T^{1/2}f$, where $f \in \mathcal{H}$ and $\langle Tf, f \rangle = 1$, is dense in the unit sphere of $\mathcal{H}_T$.

Since we assume (iii), $\|S_T^{*n}g\| \to 0$ for all $g \in \mathcal{H}_T$ of the form $g = T^{1/2}c$, $c \in \mathcal{D}$. Suppose next that $g = T^{1/2}S^k c$ for some $c \in \mathcal{D}$ and $k \geq 0$. Then for $n \geq k$,

$$S_T^{*n}g = S_T^{*n}T^{1/2}S^k c = S_T^{*n}S_T^k T^{1/2}c = S_T^{*n-k}T^{1/2}c$$

and again $\|S_T^{*n}g\| \to 0$. A routine approximation argument now shows that $\|S_T^{*n}g\| \to 0$ for every $g \in \mathcal{H}_T$, so (ii) follows.

(ii) $\Rightarrow$ (i) Let S_T be a shift operator. By the definition of S_T, for all $f \in \mathcal{H}, g \in \mathcal{H}_T$,

$$\langle T^{1/2}S_T^* g - S^* T^{1/2}g, f \rangle = \langle g, S_T T^{1/2} f \rangle - \langle g, T^{1/2}Sf \rangle = 0.$$

Hence

$$T^{1/2}S_T^* g = S^* T^{1/2}g, \qquad g \in \mathcal{H}_T,$$

and $T^{1/2}(\ker S_T^*) \subseteq \ker S^*$. Set $\mathcal{K}_T = \ker S_T^*$. Then $J = T^{1/2}|\mathcal{K}_T \in \mathcal{B}(\mathcal{K}_T, \mathcal{K})$. By the polar decomposition of an operator, $J^* = V^*R$, where $R = (JJ^*)^{1/2} \in \mathcal{B}(\mathcal{K})$ and $V \in \mathcal{B}(\mathcal{K}_T, \mathcal{K})$ is a partial isometry with initial space $\overline{J^*\mathcal{K}}$. Actually V is an isometry. For,

$$\ker V = \mathcal{K}_T \ominus J^*\mathcal{K} = \ker J \subseteq \ker T^{1/2},$$

and at the same time

$$\ker V \subseteq \mathcal{K}_T = \mathcal{H} \ominus \ker T^{1/2},$$

so $\ker V = \{0\}$.

Since S_T is a shift operator, each $g \in \mathcal{H}_T$ has a unique representation

$$g = \sum_0^\infty S_T^n k_j \quad \text{where } \{k_j\}_0^\infty \subseteq \mathcal{K}_T. \tag{3-10}$$

Set

$$Kg = \sum_0^\infty S^j V k_j. \tag{3-11}$$

This defines an isometry K that maps $\mathcal{H}_T$ into $\mathcal{H}$ and satisfies $KS_T = SK$. Define $A \in \mathcal{B}(\mathcal{H})$ by

$$Af = KT^{1/2}f, \qquad f \in \mathcal{H}.$$

For any $f \in \mathcal{H}$,

$$ASf = KT^{1/2}Sf = KS_T T^{1/2}f = SKT^{1/2}f = SAf,$$

and

$$\langle Tf, f \rangle = \langle T^{1/2}f, T^{1/2}f \rangle = \langle KT^{1/2}f, KT^{1/2}f \rangle = \langle Af, Af \rangle.$$

Thus $AS = SA$ and $T = A^*A$; that is, (i) holds.

We complete the proof by showing that the operator A constructed above is outer and $A_0 = P_0 A P_0 | \mathcal{K}$ is nonnegative. Setting $M_0 = V\mathcal{K}_T$, we obtain

$$\overline{A\mathcal{H}} = \overline{KT^{1/2}\mathcal{H}} = K\mathcal{H}_T = \sum_0^\infty \oplus S^j M_0,$$

and so $\overline{A\mathcal{H}}$ reduces S. Thus A is outer. The operator R constructed above is nonnegative. We show that $A_0 = R$. If $c \in \mathcal{K}$, then $A_0 c = P_0 A P_0 c = P_0 K T^{1/2} c$. Let $T^{1/2}c = \sum_0^\infty S_T^j k_j$ as in (3-10), so by (3-11), $KT^{1/2}c = \sum_0^\infty S^j V k_j$. Hence

$$A_0 c = P_0 K T^{1/2} c = P_0 \sum_0^\infty S^j V k_j = V k_0.$$

If P_T is the projection of $\mathcal{H}_T$ onto $\mathcal{K}_T$, then $k_0 = P_T T^{1/2} c = J^* c$, and so

$$A_0 c = V k_0 = V J^* c = Rc.$$

Therefore $A_0 = R \geq 0$, and the proof is complete.   ∎

3.5   Uniqueness Theorems

The question arises as to what extent the product A^*A determines an analytic operator $A \in \mathcal{B}(\mathcal{H})$.

THEOREM A.   *Let $A \in \mathcal{B}(\mathcal{H})$ be analytic, and let $C \in \mathcal{B}(\mathcal{H})$ be outer. Then $A^*A = C^*C$ if and only if*

$$A = BC,$$

where $B \in \mathcal{B}(\mathcal{H})$ is inner and has initial space $\overline{C\mathcal{H}}$.

Proof.   Assume that $A^*A = C^*C$. Then for any $f \in \mathcal{H}$,

$$\|Af\|^2 = \langle A^*Af, f \rangle = \langle C^*Cf, f \rangle = \|Cf\|^2.$$

Hence there is a unique partial isometry $B \in \mathcal{B}(\mathcal{H})$ with initial space $\overline{C\mathcal{H}}$ such that $A = BC$. For any $f \in \mathcal{H}$,

$$(SB - BS)Cf = SAf - ASf = 0.$$

Thus SB and BS coincide on $C\mathcal{H}$. Since C is outer, $\overline{C\mathcal{H}}$ reduces S. Therefore $g \perp C\mathcal{H}$ implies $Sg \perp C\mathcal{H}$. Since B is zero on $(C\mathcal{H})^\perp$, $SBg = 0 = BSg$ for all $g \in (C\mathcal{H})^\perp$. Hence $SB = BS$, so B is inner. This proves the necessity part of the theorem.

Conversely, let $A = BC$, where $B \in \mathcal{B}(\mathcal{H})$ is inner with initial space $\overline{C\mathcal{H}}$. Then BB^* is the projection of $\mathcal{H}$ on $\overline{C\mathcal{H}}$ by 1.2. Hence $A^*A = C^*B^*BC = C^*C$.   ∎

COROLLARY.   *Let $A, C \in \mathcal{B}(\mathcal{H})$ be two outer operators. Then $A^*A = C^*C$ if and only if*

$$A = BC,$$

where $B \in \mathcal{B}(\mathcal{H})$ is an S-constant inner operator with initial space $\overline{C\mathcal{H}}$ and final space $\overline{A\mathcal{H}}$.

Theorem B. *Let $A, C \in \mathscr{B}(\mathscr{H})$ be outer operators, and let A_0, C_0 be the diagonal entries in their matrices. If $A^*A = C^*C$, $A_0 \geq 0$, and $C_0 \geq 0$, then $A = C$.*

Lemma. *If $A \in \mathscr{B}(\mathscr{H})$ is outer, then*

$$\overline{A\mathscr{H}} = \sum_0^\infty \oplus S^j M_0(A^*), \tag{3-12}$$

where $M_0(A^*) = \overline{P_0 A \mathscr{H}}$.

The identity (3-12) is a companion to (1-8).

Proof of lemma.   Since A is outer, $\overline{A\mathscr{H}}$ reduces S. By Corollary C of 1.3,

$$\overline{A\mathscr{H}} = \sum_0^\infty \oplus S^j M,$$

where $M = P_0(\overline{A\mathscr{H}})$. Since $P_0(\overline{A\mathscr{H}})$ is closed and $P_0 A \mathscr{H} \subseteq P_0(\overline{A\mathscr{H}})$, we have $\overline{P_0 A \mathscr{H}} \subseteq P_0(\overline{A\mathscr{H}}) = M$. Conversely, if $g \in M$, then $g \in \overline{A\mathscr{H}}$ so $g = \lim_{n \to \infty} A f_n$ for some sequence $\{f_n\}_1^\infty$ in $\mathscr{H}$. Then $g = P_0 g = \lim_{n \to \infty} P_0 A f_n \in \overline{P_0 A \mathscr{H}}$. Thus $M = \overline{P_0 A \mathscr{H}}$, and the result follows.   ∎

Proof of Theorem B.   If $A^*A = C^*C$, then by the corollary to Theorem A, $A = BC$, where B is an S-constant inner operator with initial space $\overline{C\mathscr{H}}$ and final space $\overline{A\mathscr{H}}$. By Theorem B of 3.2, the diagonal entries A_0, B_0, C_0 of A, B, C satisfy $A_0 = B_0 C_0$. Hence

$$A_0^2 = C_0 B_0^* B_0 C_0 \leq C_0^2.$$

Interchanging the roles of A and C, we obtain also $C_0^2 \leq A_0^2$, and hence $A_0^2 = C_0^2$. Since A_0 and C_0 are nonnegative, $A_0 = C_0$. Since $A_0 = B_0 C_0$, B_0 coincides with the identity operator on $\overline{C_0 \mathscr{H}}$. Since $\overline{C_0 \mathscr{H}} = \overline{P_0 C \mathscr{H}} = P_0 C \mathscr{H}$, the lemma yields

$$\overline{C\mathscr{H}} = \sum_0^\infty \oplus S^j(\overline{C_0 \mathscr{H}}).$$

Thus each $f \in \overline{C\mathscr{H}}$ has the form $f = \sum_0^\infty S^j k_j$, where $k_j \in \overline{C_0 \mathscr{H}}$, $j \geq 0$. Then by (1-7),

$$Bf = \sum_0^\infty S^j B_0 k_j = \sum_0^\infty S^j k_j = f.$$

It follows that B coincides with the identity operator on $\overline{C\mathscr{H}}$. Since $A = BC$, we therefore have $A = C$.   ∎

3.6   Theorem. Inner-Outer Factorization of Toeplitz Operators

If $A \in \mathscr{B}(\mathscr{H})$ is analytic, then

$$A = BC,$$

where $C \in \mathscr{B}(\mathscr{H})$ is outer and $B \in \mathscr{B}(\mathscr{H})$ is inner with initial space $\overline{C\mathscr{H}}$. For any *such factorization,*

$$A^*A = C^*C.$$

Moreover, we may choose the factorization so that the diagonal entry C_0 in the matrix for C satisfies $C_0 \geq 0$, and then the factors B and C are unique.

Proof. Applying 3.4 to the operator $T = A^*A$, we obtain an outer operator $C \in \mathscr{B}(\mathscr{H})$ such that $A^*A = C^*C$ and $C_0 = P_0 C P_0 | \mathscr{K} \geq 0$. By Section 3.5, Theorem A, there is an inner operator $B \in \mathscr{B}(\mathscr{H})$ with initial space $\overline{C\mathscr{H}}$ such that $A = BC$.

Let $A = BC$ be any factorization where $C \in \mathscr{B}(\mathscr{H})$ is outer and $B \in \mathscr{B}(\mathscr{H})$ is inner with initial space $\overline{C\mathscr{H}}$. By 1.2, B^*B is the projection of $\mathscr{H}$ on $\overline{C\mathscr{H}}$. Therefore $A^*A = C^*B^*BC = C^*C$.

Uniqueness of the outer factor C when $C_0 \geq 0$ follows from Section 3.5, Theorem B. It remains to show that the inner factor B is also unique. If B_1, B_2 are two inner operators with initial space $\overline{C\mathscr{H}}$ such that $B_1 C = B_2 C$, then B_1 and B_2 coincide on $\overline{C\mathscr{H}}$. Since B_1 and B_2 are both zero on the orthogonal complement of $C\mathscr{H}$, $B_1 = B_2$. The result follows.   ∎

In Sections 3.7 and 3.8 we give sufficient conditions on a nonnegative Toeplitz operator $T \in \mathscr{B}(\mathscr{H})$ for the existence of an analytic operator $A \in \mathscr{B}(\mathscr{H})$ such that $T = A^*A$.

3.7   Theorem. Factorization of Nonnegative Invertible Toeplitz Operators

Let $T \in \mathscr{B}(\mathscr{H})$ be a nonnegative Toeplitz operator. If $T \geq \delta I$ for some number $\delta > 0$, then $T = A^*A$ for some analytic operator $A \in \mathscr{B}(\mathscr{H})$.

Proof. The hypotheses imply that T is invertible, $\mathscr{H}_T = \mathscr{H}$, and $S_T = T^{1/2} S T^{-1/2}$ (see Section 3.3). It follows that $S_T^{*n} \to 0$ strongly, so S_T is a shift operator, and the result follows from 3.4.   ∎

3.8   Comparison Theorem

Let $T_1, T_2 \in \mathscr{B}(\mathscr{H})$ be two nonnegative Toeplitz operators with $T_1 \leq T_2$. Assume that

(i) *$T_1 = A_1^*A_1$ for some analytic operator $A_1 \in \mathscr{B}(\mathscr{H})$, and*
(ii) *$\lim_{n \to \infty} \langle T_2 f_n, f_n \rangle = 0$ for every sequence $\{f_n\}_1^\infty$ in $\mathscr{H}$ such that $\lim_{n,k \to \infty} \langle T_2(f_n - f_k), f_n - f_k \rangle = 0$ and $\lim_{n \to \infty} \langle T_1 f_n, f_n \rangle = 0$.*

Then $T_2 = A_2^*A_2$ for some analytic operator $A_2 \in \mathscr{B}(\mathscr{H})$.

Proof. By (i) and 3.4, the Lowdenslager isometry S_{T_1} is a shift operator on $\mathscr{H}_{T_1}$. We show that S_{T_2} is a shift operator on $\mathscr{H}_{T_2}$.

Since $T_1 \leq T_2$, for each $f \in \mathscr{H}$,

$$\|T_1^{1/2}f\|^2 = \langle T_1 f, f \rangle \le \langle T_2 f, f \rangle = \|T_2^{1/2}f\|^2.$$

Hence there is a unique $C \in \mathcal{B}(\mathcal{H}_{T_2}, \mathcal{H}_{T_1})$ such that

$$CT_2^{1/2}f = T_1^{1/2}f, \quad f \in \mathcal{H}.$$

The assumption (ii) implies that $\ker C = \{0\}$, and hence the range of C^* is dense in $\mathcal{H}_{T_2}$. For each $f \in \mathcal{H}$ and $n \ge 0$,

$$CS_{T_2}^n T_2^{1/2}f = CT_2^{1/2}S^n f = T_1^{1/2}S^n f = S_{T_1}^n T_1^{1/2}f = S_{T_1}^n CT_2^{1/2}f.$$

Thus $CS_{T_2}^n = S_{T_1}^n C$ and $S_{T_2}^{*n}C^* = C^*S_{T_1}^{*n}$ for all $n = 0,1,2,\ldots$. Since S_{T_1} is a shift operator, $S_{T_1}^{*n} \to 0$ strongly on $\mathcal{H}_{T_1}$. Hence $S_{T_2}^{*n}g \to 0$ for each $g \in C^*\mathcal{H}_{T_1}$. Since the range of C^* is dense in $\mathcal{H}_{T_2}$, $S_{T_2}^{*n} \to 0$ strongly on $\mathcal{H}_{T_2}$. Therefore S_{T_2} is a shift operator, and the theorem follows by another application of 3.4. ∎

3.9 Scalar Analytic Operators

We call $V \in \mathcal{B}(\mathcal{H})$ *scalar analytic* if $AV = VA$ for every analytic operator A in $\mathcal{B}(\mathcal{H})$.

THEOREM. *An operator $V \in \mathcal{B}(\mathcal{H})$ is scalar analytic if and only if its matrix has the form (3-6), where each entry is a scalar multiple of the identity operator on $\mathcal{K}$.*

Proof. If V is scalar analytic, then V is analytic, so its matrix has the form (3-6). The matrix (3-6) must further commute with $\mathrm{diag}\{B_0, B_0, B_0, \ldots\}$ for every $B_0 \in \mathcal{B}(\mathcal{K})$, and hence the entries of (3-6) commute with all operators in $\mathcal{B}(\mathcal{K})$. Therefore the entries of (3-6) are scalar multiples of the identity operator on $\mathcal{K}$.

For the other direction it is enough to check that two matrices of the form (3-6) commute if all entries of one commute with all entries of the other. The calculation is routine. ∎

3.10 Extremal Properties of Outer Operators

We state our results here but defer proofs to the next section. We understand that whenever $A, C \in \mathcal{B}(\mathcal{H})$ are analytic, their matrices are

$$\begin{bmatrix} A_0 & 0 & 0 & \cdots \\ A_1 & A_0 & 0 & \cdots \\ A_2 & A_1 & A_0 & \cdots \\ & \cdots & \end{bmatrix}, \quad \begin{bmatrix} C_0 & 0 & 0 & \cdots \\ C_1 & C_0 & 0 & \cdots \\ C_2 & C_1 & C_0 & \cdots \\ & \cdots & \end{bmatrix}, \quad (3\text{-}13)$$

respectively, so for all $j = 0,1,2,\ldots$,

$$A_j = P_0 S^{*j} A P_0 | \mathcal{K}, \quad C_j = P_0 S^{*j} C P_0 | \mathcal{K}. \quad (3\text{-}14)$$

THEOREM A.   *Let $C \in \mathscr{B}(\mathscr{H})$ be outer, and let $V \in \mathscr{B}(\mathscr{H})$ be scalar analytic. Then*

$$\|V^*Cf\| \le \|V^*Af\| \tag{3-15}$$

for every analytic operator $A \in \mathscr{B}(\mathscr{H})$ such that $A^*A = C^*C$ and every $f \in \mathscr{H}$.

COROLLARY.   *If $C \in \mathscr{B}(\mathscr{H})$ is outer, then*

$$\sum_0^n C_j^*C_j \ge \sum_0^n A_j^*A_j, \qquad n = 0,1,2,\ldots, \tag{3-16}$$

for every analytic operator $A \in \mathscr{B}(\mathscr{H})$ such that $A^*A = C^*C$.

THEOREM B.   *Let $C \in \mathscr{B}(\mathscr{H})$ be analytic. Let $\{V_j\}_{j \in J} \subseteq \mathscr{B}(\mathscr{H})$ be scalar analytic operators, and let $\{f_k\}_{k \in K} \subseteq \mathscr{H}$ be vectors such that*

$$\|V_j^*Cf_k\| \le \|V_j^*Af_k\|, \qquad j \in J, \quad k \in K, \tag{3-17}$$

for every analytic operator $A \in \mathscr{B}(\mathscr{H})$ such that $A^*A = C^*C$. Assume that:

- (i) *The closure in the weak operator topology of the linear span of $\{V_j\}_{j \in J}$ contains S.*
- (ii) *The closed linear span of $\{P_0 f_k\}_{k \in K}$ is $\mathscr{K}$.*

Then C is outer.

THEOREM C.   *Let $C \in \mathscr{B}(\mathscr{H})$ be analytic. The following are equivalent:*

- (i) *C is outer;*
- (ii) *$C_0^*C_0 \ge A_0^*A_0$ for every analytic operator A such that $A^*A = C^*C$;*
- (iii) *for each $k \in \mathscr{K}$,*

$$\langle C_0^*C_0 k,k \rangle = \inf_{f \in \mathscr{H}} \langle C^*C(k - Sf),k - Sf \rangle. \tag{3-18}$$

Moreover, if C is outer, then (abstract Szegö infimum)

$$\left\langle \sum_0^n C_j^*C_j k,k \right\rangle = \inf_{f \in \mathscr{H}} \langle C^*C(k - S^{n+1}f),k - S^{n+1}f \rangle \tag{3-19}$$

for all $k \in \mathscr{K}$ and $n = 0,1,2,\ldots$.

3.11   Proofs

Notation is the same as in Section 3.10.

LEMMA A.   *Let $B \in \mathscr{B}(\mathscr{H})$ be a partial isometry with initial space M. If $V \in \mathscr{B}(\mathscr{H})$, $VB = BV$, and $g \in M$, then*

$$\|V^*g\| \le \|V^*Bg\| \tag{3-20}$$

with equality if and only if $BV^*g = V^*Bg$.

Proof.   By 1.2, B^*B is the projection of $\mathscr{H}$ on M. Hence $B^*Bg = g$ and

$$\|V^*g\| = \|V^*B^*Bg\| = \|B^*V^*Bg\| \le \|V^*Bg\|.$$

If equality holds, then V^*Bg is in the initial space of B^*, hence $V^*Bg = Bh$ for some $h \in M$. Then $B^*V^*Bg = B^*Bh$, $V^*B^*Bg = B^*Bh$, and so $V^*g = h$. Thus $V^*Bg = Bh = BV^*g$. Conversely, if $BV^*g = V^*Bg$, then

$$\|V^*Bg\| = \|BV^*g\| \le \|V^*g\|,$$

and so equality holds in (3-20).   ∎

LEMMA B.   *Let $B \in \mathcal{B}(\mathcal{H})$ be inner with initial space M. Let $\{V_j\}_{j \in J} \subseteq \mathcal{B}(\mathcal{H})$ with $V_j B = B V_j, j \in J$, and let $\{g_k\}_{k \in K} \subseteq M$ be vectors such that*

$$\|V_j^* g_k\| = \|V_j^* B g_k\|, \qquad j \in J, \quad k \in K. \tag{3-21}$$

Assume that:

(i) *The closure in the weak operator topology of the linear span of $\{V_j\}_{j \in J}$ contains S.*
(ii) *The closed linear span of $\{P_0 g_k\}_{k \in K}$ is $P_0 M$.*

Then B is S-constant.

Proof.   By Lemma A,

$$B V_j^* g_k = V_j^* B g_k, \qquad j \in J, \quad k \in K.$$

Hence by (i),

$$B S^* g_k = S^* B g_k, \qquad k \in K.$$

Act on both sides with S and use $P_0 = I - SS^*$ to get

$$B P_0 g_k = P_0 B g_k, \qquad k \in K.$$

Hence by (ii), $B\mathcal{K} \subseteq \mathcal{K}$, and so B has a diagonal matrix (Section 3.2, Corollary to Theorem C). Therefore B is S-constant.   ∎

Proof of 3.10, Theorem A.   If $A \in \mathcal{B}(\mathcal{H})$ is analytic with $A^*A = C^*C$, then by Section 3.5, Theorem A, $A = BC$ for some inner operator $B \in \mathcal{B}(\mathcal{H})$ with initial space $\overline{C\mathcal{H}}$. Applying Lemma A with $g = Cf$, we obtain (3-15) from (3-20).   ∎

Proof of 3.10, corollary to Theorem A.   Let $A \in \mathcal{B}(\mathcal{H})$ be analytic with $A^*A = C^*C$. In the theorem choose $V = S^{n+1}$ for a fixed $n \ge 0$. Set $f = P_0 h$ for an arbitrary $h \in \mathcal{H}$. Then (3-15) yields

$$\|S^{*n+1} C P_0 h\| \le \|S^{*n+1} A P_0 h\|.$$

By the arbitrariness of h,

$$P_0 C^* S^{n+1} S^{*n+1} C P_0 \le P_0 A^* S^{n+1} S^{*n+1} A P_0.$$

Since $P_0 = I - SS^*$, we get

$$P_0 C^*(I - \sum_0^n S^j P_0 S^{*j}) C P_0 \le P_0 A^*(I - \sum_0^n S^j P_0 S^{*j}) A P_0.$$

Since $C^*C = A^*A$, this is the same as

$$\sum_0^n P_0 C^* S^j P_0 S^{*j} C P_0 \ge \sum_0^n P_0 A^* S^j P_0 S^{*j} A P_0,$$

which by (3-14) is (3-16).   ∎

Proof of 3.10, Theorem B.   By 3.6 there is a factorization $C = BA$, where A is outer, B is inner with initial space $\overline{A\mathcal{H}}$, and $A^*A = C^*C$. Claim: B is S-constant. We apply Lemma B with $g_k = Af_k$, $k \in K$. By (3-17), for all $j \in J$, $k \in K$,

$$\|V_j^*Bg_k\| = \|V_j^*Cf_k\| \le \|V_j^*Af_k\| = \|V_j^*g_k\|.$$

Since the reverse inequality is automatic by Lemma A, (3-21) holds. The only hypothesis in Lemma B that is not clearly met is that the closed linear span of the vectors P_0g_k, $k \in K$, is P_0M, where $M = \overline{A\mathcal{H}}$ is the initial space of B. To see this, note that $P_0A = P_0AP_0$, and hence $P_0g_k = P_0Af_k = P_0AP_0f_k$, $k \in K$. Since by hypothesis the vectors P_0f_k, $k \in K$, span a dense subset of $\mathcal{K}$, we have

$$\bigvee_{k \in K} \{P_0g_k\} = \overline{P_0A\mathcal{K}} = \overline{P_0A\mathcal{H}} = P_0M$$

(the last equality is by 3.5, Lemma to Theorem B). The hypotheses of Lemma B are thus satisfied. By Lemma B, B is S-constant.

Since B is S-constant and A is outer, $C = BA$ is also outer. For, the final space $\overline{C\mathcal{H}}$ of B reduces S by Section 1.7, Theorem C.   ∎

Proof of 3.10, Theorem C.   (i) $\Rightarrow$ (iii) Let C be outer; so $\overline{C\mathcal{H}}$ reduces S. Fix $k \in \mathcal{K}$ and $n \ge 0$. By Section 1.2 (Corollary 2), the infimum of $\|Ck - S^{n+1}g\|$ over all $g \in \overline{C\mathcal{H}}$ is attained with $g = S^{*n+1}Ck$. Hence

$$\inf_{f \in \mathcal{H}} \langle C^*C(k - S^{n+1}f), k - S^{n+1}f \rangle$$

$$= \inf_{f \in \mathcal{H}} \|Ck - S^{n+1}Cf\|^2$$

$$= \inf_{g \in \overline{C\mathcal{H}}} \|Ck - S^{n+1}g\|^2 = \|Ck - S^{n+1}S^{*n+1}Ck\|^2$$

$$= \left\| \sum_0^n S^j(I - SS^*)S^{*j}Ck \right\|^2 = \sum_0^n \|P_0S^{*j}Ck\|^2$$

$$= \sum_0^n \|C_jk\|^2 = \left\langle \sum_0^n C_j^*C_jk, k \right\rangle.$$

Thus (3-19) holds and (iii) follows.

(iii) $\Rightarrow$ (ii) Assume (iii), and let A be analytic with $A^*A = C^*C$. Then for any $k \in \mathcal{K}$,

$$\langle C_0^*C_0k, k \rangle = \inf_{f \in \mathcal{H}} \langle A^*A(k - Sf), k - Sf \rangle$$

$$= \inf_{f \in \mathcal{H}} \|Ak - SAf\|^2$$

$$\ge \inf_{g \in \mathcal{H}} \|Ak - Sg\|^2$$

$$= \langle A_0^*A_0k, k \rangle.$$

Hence (ii) holds.

(ii) $\Rightarrow$ (i) Assume (ii). By 3.6, $C = BA$, where A is outer and B is inner with initial space $\overline{A\mathcal{H}}$, and $C^*C = A^*A$. By (ii), $C_0^*C_0 \ge A_0^*A_0$. Since $C = BA$, $C_0 = B_0A_0$, where $B_0 = P_0BP_0|\mathcal{K}$ satisfies $\|B_0\| \le 1$. Hence $C_0^*C_0 \le A_0^*A_0$, so $C_0^*C_0 =$

$A_0^* A_0$ and B_0 is isometric on $\overline{A_0 \mathcal{K}}$. For any $k \in \mathcal{K}$,

$$\|A_0 k\| = \|B_0 A_0 k\| = \|P_0 B A_0 k\| \le \|B A_0 k\| \le \|A_0 k\|.$$

Therefore equality holds throughout, and so

$$B_0 A_0 k = P_0 B A_0 k = B A_0 k.$$

Thus B and B_0 coincide on $\overline{A_0 \mathcal{K}}$. Then BS^j and $S^j B_0$ coincide on $\overline{A_0 \mathcal{K}}$ for any $j \ge 0$. Hence for any $j \ge 0$,

$$BS^j \overline{A_0 \mathcal{K}} = S^j B_0 \overline{A_0 \mathcal{K}} = S^j \overline{C_0 \mathcal{K}}.$$

By Section 3.5 (see the Lemma to Theorem B),

$$\overline{C \mathcal{H}} = B \overline{A \mathcal{H}} = B \left(\sum_0^\infty \oplus \, S^j \overline{A_0 \mathcal{K}} \right) = \sum_0^\infty \oplus \, S^j \overline{C_0 \mathcal{K}}.$$

It follows that $\overline{C \mathcal{H}}$ reduces S; that is, (i) holds. This completes the proof. ■

Examples and Addenda

1. *The "natural" Toeplitz operators on* l^2, $H^2(\Gamma)$, $L^2(0, \infty)$, *and* $H^2(R)$. Define

$$S_1 : (c_0, c_1, c_2, \ldots) \to (0, c_0, c_1, \ldots), \qquad (c_0, c_1, c_2, \ldots) \in \mathcal{H}_1 = l^2,$$

$$S_2 : f(e^{it}) \to e^{it} f(e^{it}), \qquad\qquad\qquad\qquad f \in \mathcal{H}_2 = H^2(\Gamma),$$

$$S_3 : f(x) \to f(x) - \int_0^x e^{-\frac{1}{2}(x-t)} f(t)\, dt, \qquad f \in \mathcal{H}_3 = L^2(0, \infty),$$

$$S_4 : F(x) \to \frac{x - \frac{1}{2}i}{x + \frac{1}{2}i} F(x), \qquad\qquad\qquad F \in \mathcal{H}_4 = H^2(R).$$

The operator S_3 is the Laguerre shift (Examples and Addenda to Chapter 1, no. 6). Associated with these operators are the orthonormal bases

$$\{\phi_1^{(n)}\}_0^\infty = \{(1, 0, 0, \ldots), (0, 1, 0, \ldots), \ldots\} \subseteq \mathcal{H}_1,$$

$$\{\phi_2^{(n)}\}_0^\infty = \{e^{int}\}_0^\infty \subseteq \mathcal{H}_2,$$

$$\{\phi_3^{(n)}\}_0^\infty = \{e^{-\frac{1}{2}t} L_n(t)\}_0^\infty \subseteq \mathcal{H}_3,$$

$$\{\phi_4^{(n)}\}_0^\infty = \left\{ \frac{i(2\pi)^{-1/2}}{x + \frac{1}{2}i} \left(\frac{x - \frac{1}{2}i}{x + \frac{1}{2}i} \right)^n \right\}_0^\infty \subseteq \mathcal{H}_4.$$

For each $j = 1, 2, 3, 4$, S_j is a shift operator on $\mathcal{H}_j$ and

$$S_j : \phi_j^{(n)} \to \phi_j^{(n+1)}, \qquad n = 0, 1, 2, \ldots.$$

These operators are unitarily equivalent by means of the isomorphisms $U_{jk} \in \mathcal{B}(\mathcal{H}_j, \mathcal{H}_k)$ such that

$$U_{jk} : \phi_j^{(n)} \to \phi_k^{(n)}, \qquad n = 0, 1, 2, \ldots.$$

The isomorphism U_{34} is the Paley-Wiener representation of $H^2(R)$. That is, $U_{34} = \mathscr{F}^{-1}|L^2(0,\infty)$, where

$$\mathscr{F}: f(x) \to \underset{T \to \infty}{\text{l.i.m.}} \frac{1}{\sqrt{2\pi}} \int_{-T}^{T} e^{-ixt} f(t)\, dt$$

is the Fourier-Plancherel operator on $L^2(-\infty,\infty)$ [$L^2(0,\infty)$ is viewed as a subspace of $L^2(-\infty,\infty)$]. See Dym and McKean [1972], p. 162.

S_1-Toeplitz Operators.   By 3.1, a bounded operator T_1 on $\mathscr{H}_1 = l^2$ is S_1-Toeplitz if and only if its matrix relative to the canonical basis is a Toeplitz matrix, that is, it has the form $[c_{j-k}]_{j,k=0}^{\infty}$ for some sequence $\{c_n\}_{-\infty}^{\infty}$.

A Toeplitz matrix $[c_{j-k}]_{j,k=0}^{\infty}$ induces a bounded operator T_1 on $\mathscr{H}_1$ if and only if

$$c_n = \int_{\Gamma} w(e^{it}) e^{-int}\, d\sigma(e^{it}), \qquad n = 0, \pm 1, \pm 2, \dots, \tag{3-22}$$

for some $w \in L^{\infty}(\sigma)$. In this case $\|T_1\| = \|w\|_{\infty}$ and T_1 is S_1-Toeplitz.

S_2-Toeplitz Operators.   Let P be the projection of $L^2(\sigma)$ on $H^2(\Gamma)$. If $w \in L^{\infty}(\sigma)$, the operator on $\mathscr{H}_2$ defined by

$$T_2 f = Pwf, \qquad f \in \mathscr{H}_2 = H^2(\Gamma),$$

is S_2-Toeplitz, and $\|T_2\| = \|w\|_{\infty}$. Every S_2-Toeplitz operator has this form (see 6.2).

S_3-Toeplitz Operators.   Let P_+ be the projection of $L^2(-\infty,\infty)$ on $L^2(0,\infty)$. If $W \in L^{\infty}(-\infty,\infty)$, the operator on $\mathscr{H}_3$ defined by

$$T_3 f = P_+ \mathscr{F} W \mathscr{F}^{-1} f, \qquad f \in \mathscr{H}_3 = L^2(0,\infty), \tag{3-23}$$

is S_3-Toeplitz, and $\|T_3\| = \|W\|_{\infty}$. Every S_3-Toeplitz operator has this form.

S_4-Toeplitz Operators.   Let Q be the projection of $L^2(-\infty,\infty)$ on $H^2(R)$. If $W \in L^{\infty}(-\infty,\infty)$, the operator on $\mathscr{H}_4$ defined by

$$T_4 f = QWf, \qquad f \in \mathscr{H}_4 = H^2(R),$$

is S_4-Toeplitz, and $\|T_4\| = \|W\|_{\infty}$. Every S_4-Toeplitz operator has this form.

ISOMORPHISM THEOREM (*Rosenblum* [1965], *Devinatz* [1967]).   *Let $w \in L^{\infty}(\sigma)$ and $W \in L^{\infty}(-\infty,\infty)$ be related by*

$$W(x) = w\left(\frac{x - \frac{1}{2}i}{x + \frac{1}{2}i}\right), \qquad x \text{ real},$$

and define $\{c_n\}_{-\infty}^{\infty}$ by (3-22). Define operators T_1, T_2, T_3, T_4 as above. Then for each $j = 1,2,3,4$,

$$c_{m-n} = \langle T_j \phi_j^{(n)}, \phi_j^{(m)} \rangle_{\mathscr{H}_j}, \qquad m,n = 0,1,2,\dots.$$

Therefore the operators T_1, T_2, T_3, T_4 are unitarily equivalent by means of the isomorphisms $U_{jk}, j,k = 1,2,3,4$.

Proof. By the definition of T_1 and (3-22), for any $m,n = 0,1,2,\ldots,$

$$\langle T_1 \phi_1^{(n)}, \phi_1^{(m)} \rangle_{\mathcal{H}_1} = c_{m-n}$$

$$= \int_\Gamma w(e^{it}) e^{it(n-m)} \, d\sigma(e^{it})$$

$$= \langle T_2 \phi_2^{(n)}, \phi_2^{(m)} \rangle_{\mathcal{H}_2}.$$

Changing variables with the substitution $e^{it} = (x - \frac{1}{2}i)/(x + \frac{1}{2}i)$, we obtain also

$$c_{m-n} = \int_\Gamma w(e^{it}) e^{it(n-m)} \, d\sigma(e^{it})$$

$$= \frac{1}{2\pi} \int_{-\infty}^{\infty} \frac{W(x)}{x^2 + \frac{1}{4}} \left(\frac{x - \frac{1}{2}i}{x + \frac{1}{2}i} \right)^{n-m} dx$$

$$= \langle T_4 \phi_4^{(n)}, \phi_4^{(m)} \rangle_{\mathcal{H}_4}$$

$$= \langle QW\mathscr{F}^{-1}\phi_3^{(n)}, \mathscr{F}^{-1}\phi_3^{(m)} \rangle_{\mathcal{H}_4}$$

$$= \langle \mathscr{F}Q\mathscr{F}^{-1}\mathscr{F}W\mathscr{F}^{-1}\phi_3^{(n)}, \phi_3^{(m)} \rangle_{\mathcal{H}_3}$$

$$= \langle P_+ \mathscr{F}W\mathscr{F}^{-1}\phi_3^{(n)}, \phi_3^{(m)} \rangle_{\mathcal{H}_3}$$

$$= \langle T_3 \phi_3^{(n)}, \phi_3^{(m)} \rangle_{\mathcal{H}_3}.$$

For the next to last equality we used the relation $Q = \mathscr{F}^{-1}P\mathscr{F}$, which is a consequence of the Paley-Wiener representation of $H^2(R)$. The result follows. ∎

Wiener-Hopf Operators. If $K \in L^1(-\infty,\infty)$, then the operator

$$T: f(x) \to \int_0^\infty K(x - t)f(t) \, dt$$

is everywhere defined and bounded on $L^2(0,\infty)$. In fact, T has the form (3-23), where

$$W(x) = \int_{-\infty}^\infty e^{ixt} K(t) \, dt.$$

In particular, T is an S_3-Toeplitz operator. An operator of this form is called a *Wiener-Hopf operator*. See Kreĭn [1958] and Gohberg and Kreĭn [1958].

Examples. (i) For $w(e^{it}) = \cos t = \frac{1}{2}(e^{it} + e^{-it})$, we have

$$c_n = \begin{cases} \frac{1}{2} & \text{if } n = \pm 1, \\ 0 & \text{otherwise,} \end{cases}$$

and

$$W(x) = w\left(\frac{x - \frac{1}{2}i}{x + \frac{1}{2}i}\right) = \frac{x^2 - \frac{1}{4}}{x^2 + \frac{1}{4}} = 1 - \frac{1}{2} \int_{-\infty}^{\infty} e^{ixt} e^{-\frac{1}{2}|t|} dt.$$

Thus

$$T_1 \sim \begin{bmatrix} 0 & \frac{1}{2} & 0 & 0 & \cdots \\ \frac{1}{2} & 0 & \frac{1}{2} & 0 & \cdots \\ 0 & \frac{1}{2} & 0 & \frac{1}{2} & \cdots \\ & & \cdots & & \end{bmatrix} \qquad \text{on } \mathscr{H}_1 = l^2,$$

$$T_2 f = P(\cos t) f, \qquad\qquad f \in \mathscr{H}_2 = H^2(\Gamma),$$

$$T_3 : f(x) \to f(x) - \frac{1}{2} \int_0^{\infty} e^{-\frac{1}{2}|x-t|} f(t)\, dt, \qquad f \in \mathscr{H}_3 = L^2(0,\infty),$$

$$T_4 f = P_+ \frac{x^2 - \frac{1}{4}}{x^2 + \frac{1}{4}} f, \qquad\qquad f \in \mathscr{H}_4 = H^2(R).$$

(ii) For any complex parameter ρ, $|\rho| < 1$, set

$$w(e^{it}) = \frac{1 - \rho^2}{1 - 2\rho \cos t + \rho^2}.$$

Then $c_n = \rho^{|n|}$, $n = 0, \pm 1, \pm 2, \ldots$. Setting $\mu = (1 + \rho)/(1 - \rho)$, we find that $\operatorname{Re} \mu > 0$ and

$$W(x) = w\left(\frac{x - \frac{1}{2}i}{x + \frac{1}{2}i}\right) = \mu \frac{x^2 + \frac{1}{4}}{x^2 + \frac{1}{4}\mu^2} = \mu - \frac{1}{4}(\mu^2 - 1) \int_{-\infty}^{\infty} e^{ixt} e^{-\frac{1}{2}\mu|t|} dt.$$

Thus

$$T_1 \sim \begin{bmatrix} 1 & \rho & \rho^2 & \rho^3 & \cdots \\ \rho & 1 & \rho & \rho^2 & \cdots \\ \rho^2 & \rho & 1 & \rho & \cdots \\ & & \cdots & & \end{bmatrix} \qquad \text{on } \mathscr{H}_1 = l^2,$$

$$T_2 f = P \frac{1 - \rho^2}{1 - 2\rho \cos t + \rho^2} f, \qquad\qquad f \in \mathscr{H}_2 = H^2(\Gamma),$$

$$T_3 : f(x) \to \mu f(x) - \frac{1}{4}(\mu^2 - 1) \int_0^{\infty} e^{-\frac{1}{2}\mu|x-t|} f(t)\, dt, \qquad f \in \mathscr{H}_3 = L^2(0, \infty),$$

$$T_4 f = P_+ \mu \frac{x^2 + \frac{1}{4}}{x^2 + \frac{1}{4}\mu^2} f, \qquad\qquad f \in \mathscr{H}_4 = H^2(R).$$

Any number of similar examples can in principle be constructed. However, it is typically the case that an operator is simple and "natural" in one scheme and complicated or unrecognizable in another. We invite the reader to try the example

$$T: f(x) \to \int_0^\infty K(x - t)f(t)\,dt, \quad f \in L^2(0, \infty), \quad \text{where } K(x) = \int_{|x|}^\infty y^{-1}e^{-y}\,dy.$$

This operator is related to Milne's equation. See Dym and McKean [1972], pp. 176–178, 181–184.

2. *Abstract Szegö infimum.* Let $T \in \mathcal{B}(\mathcal{H})$ be a nonnegative Toeplitz operator relative to some shift operator S. If $k \in \mathcal{K} = \ker S^*$, then

$$\inf_{f \in \mathcal{H}} \langle T(k - Sf), k - Sf \rangle > 0$$

if and only if k has a nonzero projection on $(\mathcal{K} \cap T^{1/2}\mathcal{H})^-$ (Moore [1972a]; see also Moore [1972b], Moore, Rosenblum, and Rovnyak [1975]).

Concrete Spectral Theory

In nos. 3–9 below we sketch the explicit diagonalization of a self-adjoint Toeplitz operator when the underlying shift has multiplicity 1.

3. *Notation and preliminaries.* Let S be multiplication by $e^{i\theta}$ on $H^2(\Gamma)$, and let P be the projection of $L^2(\sigma)$ on $H^2(\Gamma)$. For any $w \in L^\infty(\sigma)$ define $T(w)$ on $H^2(\Gamma)$ by

$$T(w)f = Pwf, \quad f \in H^2(\Gamma).$$

By no. 1 above, $T(w)$ is S-Toeplitz and $\|T(w)\| = \|w\|_\infty$. Every S-Toeplitz operator has this form. Moreover:

(i) $T(w)$ is self-adjoint if and only if w is essentially real valued;
(ii) $T(w) \geq 0$ if and only if $w \geq 0$ σ-a.e.;
(iii) if $a \in H^\infty(\Gamma)$, then $T(w) = T(a)^*T(a)$ if and only if $w = |a|^2$ σ-a.e. .

4. Let w be a real valued function in $L^\infty(\sigma)$. Then $T(w)$ is self-adjoint with spectrum $\mathrm{sp}(T(w)) = [c,d]$, where $c = \mathrm{ess\,inf}\,w$ and $d = \mathrm{ess\,sup}\,w$. If w is not equal σ-a.e. to a constant, then $T(w)$ has no point spectrum (Hartman and Wintner [1950], [1954]; see also Douglas [1972], Putnam [1967], Widom [1965]).

5. *Generating formula for resolvents* (Kreĭn [1958], Calderón, Spitzer and Widom [1959], Rosenblum [1960]).

THEOREM. *Let w be a real valued function in $L^\infty(\sigma)$, and set $c = \mathrm{ess\,inf}\,w$. For any $\alpha, \beta \in D$ and $z \in \mathbf{C}\backslash[c,\infty)$,*

$$\langle (T(w) - zI)^{-1}(1 - \bar{\alpha}e^{it})^{-1}, (1 - \bar{\beta}e^{it})^{-1}\rangle_2 = \bar{a}(\alpha,\bar{z})a(\beta,z)/(1 - \bar{\alpha}\beta), \quad (3\text{-}24)$$

where for $\alpha \in D$ *and* $z \in \mathbf{C}\backslash[c,\infty)$,

$$a(\alpha,z) = \exp\left(-\frac{1}{2}\int_\Gamma \frac{e^{it}+\alpha}{e^{it}-\alpha}\log\left[w(e^{it})-z\right]d\sigma\right) \tag{3-25}$$

with the principal branch of the logarithm.

LEMMA. *If* $w \geq \delta$ σ-*a.e. for some* $\delta > 0$, *then* $T(w)$ *is invertible and* $T(w)^{-1} = T(a)T(a)^*$, *where* $a \in H^\infty(\Gamma)$ *is any outer function such that* $1/w = |a|^2$ σ-*a.e.*

Proof of lemma.   By no. 3(iii), $T(w) = T(1/a)^*T(1/a)$.   ∎

Proof of theorem.   By analyticity it is enough to prove (3-24) for $z = x$, $x < c$. By the lemma, for such x,

$$(T(w) - xI)^{-1} = T(a(\cdot,x))T(a(\cdot,x))^*,$$

and (3-24) follows by a straightforward calculation.   ∎

6. *Absolute continuity.*   Let T be a bounded self-adjoint operator on a Hilbert space $\mathscr{H}$ with spectral representation $T = \int t\, dE(t)$. One can view $E(\cdot)$ either as a spectral measure $\Delta \to E(\Delta)$ on real Borel sets (Halmos [1951]) or as a resolution of the identity $t \to E(t)$ on $(-\infty,\infty)$ (Stone [1932], Riesz and Sz.-Nagy [1955]). There is a unique decomposition $\mathscr{H} = \mathscr{H}_{ac} \oplus \mathscr{H}_s$, where $\mathscr{H}_{ac}$, $\mathscr{H}_s$ are reducing subspaces for T such that (i) for all $u \in \mathscr{H}_{ac}$, the Borel measure $\Delta \to \langle E(\Delta)u,u\rangle$ is absolutely continuous with respect to Lebesgue measure, and (ii) for all $v \in \mathscr{H}_s$, the Borel measure $\Delta \to \langle E(\Delta)v,v\rangle$ is singular with respect to Lebesgue measure (Kato [1976], pp. 518–521). We call T *absolutely continuous* if $\mathscr{H} = \mathscr{H}_{ac}$, that is, $\mathscr{H}_s = \{0\}$.

THEOREM (*Rosenblum* [1960]).   *Let* w *be a real valued function in* $L^\infty(\sigma)$. *If* w *is not equal* σ-*a.e. to a constant, then* $T(w)$ *is absolutely continuous.*

The original proof was simplified by I.I. Hirschman, Jr. (private communication), whose argument runs as follows.

LEMMA A.   *For almost all real* x,

$$\int_\Gamma |\log|w(e^{i\theta}) - x||\,d\sigma < \infty. \tag{3-26}$$

Proof.   For $t > \|w\|_\infty$ and $e^{i\theta} \in \Gamma$,

$$\int_{-t}^{t} \log|w(e^{i\theta}) - x|\,dx = \int_0^{t-w(e^{i\theta})} \log y\,dy + \int_0^{t+w(e^{i\theta})} \log y\,dy$$

$$= [t - w(e^{i\theta})]\log[t - w(e^{i\theta})]$$

$$+ [t + w(e^{i\theta})]\log[t + w(e^{i\theta})] - 2t$$

$$\geq K_t,$$

where K_t is a constant, $K_t > -\infty$. Hence

$$\int_\Gamma \int_{-t}^t \log|w(e^{i\theta}) - x|\, dx\, d\sigma > -\infty,$$

and the result follows.   ∎

Lemma B.   *If* ess inf $w < s < t <$ ess sup w *and* $\alpha \in D$, *then*

$$\int_s^t \exp\left(-\int_\Gamma P(\alpha,e^{i\theta})\log|w(e^{i\theta}) - x|\, d\sigma\right) dx < \infty. \tag{3-27}$$

Proof.   Define $a(\alpha,z)$ by (3-25). The function $V_\alpha(z) = \mathrm{Im}\,\bar{a}(\alpha,\bar{z})a(\alpha,z)$ is positive and harmonic on Π, so (Appendix, Section 6),

$$\int_{-\infty}^\infty V_\alpha(x)(1 + x^2)^{-1}\, dx < \infty. \tag{3-28}$$

For any fixed $e^{i\theta} \in \Gamma$,

$$\lim_{y\downarrow 0} \log[w(e^{i\theta}) - x - iy] = \log|w(e^{i\theta}) - x| - i\pi\chi_{\gamma(x)}(e^{i\theta}),$$

where $\gamma(x) = \{e^{it}: w(e^{it}) < x\}$. Thus

$$V_\alpha(x) = \lim_{y\downarrow 0} \mathrm{Im}\,\exp\left(-\frac{1}{2}\int_\Gamma \frac{e^{-i\theta} + \bar{\alpha}}{e^{-i\theta} - \bar{\alpha}}\log[w(e^{i\theta}) - x - iy]\, d\sigma\right.$$

$$\left.-\frac{1}{2}\int_\Gamma \frac{e^{i\theta} + \alpha}{e^{i\theta} - \alpha}\log[w(e^{i\theta}) - x - iy]\, d\sigma\right)$$

$$= \mathrm{Im}\,\exp\left(-\int_\Gamma P(\alpha,e^{i\theta})[\log|w(e^{i\theta}) - x| - i\pi\chi_{\gamma(x)}(e^{i\theta})]\, d\sigma\right)$$

$$= \exp\left(-\int_\Gamma P(\alpha,e^{i\theta})\log|w(e^{i\theta}) - x|\, d\sigma\right)\sin\left(\pi\int_{\gamma(x)} P(\alpha,e^{i\theta})\, d\sigma\right).$$

The second equality follows from the Lebesgue dominated convergence theorem, by Lemma A and the elementary inequality $|\log|z|| \le |\log|x|| + |z|$. The sine factor is bounded away from 0 for $s \le x \le t$, so (3-27) follows from (3-28).   ∎

Proof of theorem.   Let $T(w) = \int t\, dE(t)$. For $\alpha \in D$ set $k_\alpha(e^{i\theta}) = (1 - \bar{\alpha}e^{i\theta})^{-1}$. By no. 4 above, the function $t \to \langle E(t)k_\alpha,k_\alpha\rangle_2$ is continuous on $(-\infty,\infty)$ and constant on $(-\infty,c]$, $[d,\infty)$ ($c =$ ess inf w, $d =$ ess sup w). By (3-24), for

$z \in \mathbf{C} \backslash [c, \infty)$,

$$\frac{y}{\pi} \int_{-\infty}^{\infty} \frac{d \langle E(t) k_\alpha, k_\alpha \rangle_2}{(t - x)^2 + y^2}$$

$$= \frac{1}{\pi} \operatorname{Im} \langle [T(w) - zI]^{-1} k_\alpha, k_\alpha \rangle_2$$

$$= \pi^{-1} (1 - |\alpha|^2)^{-1} \operatorname{Im} \bar{a}(\alpha, \bar{z}) a(\alpha, z)$$

$$= \pi^{-1} (1 - |\alpha|^2)^{-1} \operatorname{Im} \exp \left(- \int_\Gamma P(\alpha, e^{i\theta}) \log [w(e^{i\theta}) - z] \, d\sigma \right)$$

$$= \pi^{-1} (1 - |\alpha|^2)^{-1} \exp \left(- \int_\Gamma P(\alpha, e^{i\theta}) \log |w(e^{i\theta}) - z| \, d\sigma \right)$$

$$\cdot \sin \left(- \int_\Gamma P(\alpha, e^{i\theta}) \arg [w(e^{i\theta}) - z] \, d\sigma \right)$$

$$\leq \pi^{-1} (1 - |\alpha|^2)^{-1} \exp \left(- \int_\Gamma P(\alpha, e^{i\theta}) \log |w(e^{i\theta}) - x| \, d\sigma \right).$$

By Lemma B and the Stieltjes inversion formula (Appendix, Section 6), the function $t \to \langle E(t) k_\alpha, k_\alpha \rangle_2$ is absolutely continuous on (c, d). Hence k_α belongs to the absolutely continuous subspace $\mathscr{H}_{ac}$ for $T(w)$. Since $\alpha \in D$ is arbitrary, $\mathscr{H}_{ac} = H^2(\Gamma)$ and the result follows.   ■

7. *Multiplicity theory.* Let T be a bounded self-adjoint operator on a separable Hilbert space $\mathscr{H}$ with spectrum $\operatorname{sp}(T)$.

THEOREM A.   *There is a direct integral Hilbert space*

$$\mathscr{K} = \int_X \oplus \mathscr{K}(x) \, d\mu, \tag{3-29}$$

where $X = \operatorname{sp}(T)$, such that T is unitarily equivalent to multiplication by x on $\mathscr{K}$.

We shall not prove this theorem, but we include the definition of the space (3-29). In (3-29), X is a compact subset of $(-\infty, \infty)$, and μ is a finite nonnegative Borel measure on X. For μ-a.e. $x \in X$, $\mathscr{K}(x)$ is a separable Hilbert space. A class $\mathscr{M}$ of "measurable" functions is assumed given such that:

(M1) The elements of $\mathscr{M}$ are functions f on X such that $f(x) \in \mathscr{K}(x)$ μ-a.e.
(M2) For any $f, g \in \mathscr{M}$, the scalar valued function $x \to \langle f(x), g(x) \rangle_{\mathscr{K}(x)}$ is μ-measurable.
(M3) If g is a function on X such that $g(x) \in \mathscr{K}(x)$ μ-a.e. and $x \to \langle f(x), g(x) \rangle_{\mathscr{K}(x)}$ is μ-measurable for all $f \in \mathscr{M}$, then $g \in \mathscr{M}$.

(M4) There is a sequence $\{p_j\}_1^\infty \subseteq \mathcal{M}$ such that $\mathcal{K}(x) = \bigvee\{p_j(x): j = 1,2,\ldots\}$ μ-a.e.

Functions that are equal μ-a.e. are identified. The Hilbert space (3-29) is defined as the space of all $f \in \mathcal{M}$ such that $\int_X \|f(x)\|^2_{\mathcal{K}(x)} d\mu < \infty$ in the inner product

$$\langle f,g \rangle = \int_X \langle f(x),g(x) \rangle_{\mathcal{K}(x)} d\mu.$$

In order to define the space (3-29), it is thus necessary to specify a class $\mathcal{M}$ of functions satisfying (M1)–(M4). In applications, the following result is helpful for this purpose.

LEMMA.   *Let μ be a finite nonnegative Borel measure on a compact set $X \subseteq (-\infty,\infty)$. For μ-a.e. $x \in X$, let $\mathcal{K}(x)$ be a separable Hilbert space. Assume given a sequence $\{q_j\}_1^\infty$ of functions on X such that $q_j(x) \in \mathcal{K}(x)$ μ-a.e. for each $j \geq 1$, and*

   (i) *for each $j,k \geq 1$, $x \to \langle q_j(x),q_k(x) \rangle_{\mathcal{K}(x)}$ is μ-measurable, and*
   (ii) *$\bigvee\{q_j(x): j = 1,2,\ldots\} = \mathcal{K}(x)$ μ-a.e.*

Define $\mathcal{M}$ to be the class of all functions f on X such that $f(x) \in \mathcal{K}(x)$ μ-a.e. and $x \to \langle f(x),q_j(x) \rangle_{\mathcal{K}(x)}$ is μ-measurable for each $j \geq 1$. Then $\mathcal{M}$ satisfies (M1)–(M4), and $\mathcal{M}$ is the only such class containing $\{q_j\}_1^\infty$.

See Dixmier [1969], p. 145.

In the situation of Theorem A, we define $m(x) = \dim \mathcal{K}(x)$ μ-a.e. on $\mathrm{sp}(T)$. We call m a *multiplicity function* for T. The quantities $(\mathrm{sp}(T),m,\mu)$ are called the *unitary invariants* for T. The terminology is justified by the following result.

THEOREM B.   *Let T_j be a bounded self-adjoint operator on a separable Hilbert space $\mathcal{H}_j$ with associated triple $(\mathrm{sp}(T_j),\mu_j,m_j)$ as above, $j = 1,2$. Then T_1 and T_2 are unitarily equivalent if and only if (i) $\mathrm{sp}(T_1) = \mathrm{sp}(T_2)$, (ii) μ_1 and μ_2 are mutually absolutely continuous, that is, they have the same class of null sets, and (iii) $m_1 = m_2$ μ_j-a.e. ($j = 1,2$).*

See Dixmier [1969] or Nielsen [1980]. The complete diagonalization of a given self-adjoint operator T may thus be achieved by exhibiting the associated triple $(\mathrm{sp}(T),\mu,m)$ of unitary invariants.

8. *Diagonalization of a self-adjoint Toeplitz operator.*   The *index* of a Borel set $E \subseteq \Gamma$ is defined to be (α) 0 if $E = \varnothing$ or Γ modulo a σ-null set, (β) $n = 1,2,3,\ldots$ if E is the disjoint union of n closed arcs $A_1,\ldots,A_n$ such that $0 < \sigma(A_j) < 1$, $j = 1,\ldots,n$, modulo a σ-null set, and (γ) ∞ otherwise.

THEOREM (*Ismagilov* [1963], *Rosenblum* [1965]).   *Let w be a real valued function in $L^\infty(\sigma)$ that is not equal σ-a.e. to a constant. The unitary invariants $(\mathrm{sp}(T),m,\mu)$ for $T = T(w)$ are given by*

   (i) *$\mathrm{sp}(T) = [c,d]$, where $c = $ ess inf w, $d = $ ess sup w,*
   (ii) *$m(x) = $ index of $E_x = \{e^{it}: w(e^{it}) \geq x\}$ a.e. on $\mathrm{sp}(T)$, and*
   (iii) *$\mu = $ Lebesgue measure on $\mathrm{sp}(T)$.*

LEMMA. *Let $E \subseteq \Gamma$ be a Borel set, and let*

$$\phi(z) = \frac{1}{2\pi i} \exp\left(\pi i \int_E \frac{e^{it} + z}{e^{it} - z} \, d\sigma \right), \qquad z \in D.$$

Then $\operatorname{Re} \phi(z) > 0$ on D, and $\operatorname{Re} \phi(e^{i\theta}) = 0$ σ-a.e. on Γ. There is a nonnegative singular Borel measure v on Γ such that

$$\frac{\phi(\beta) + \bar{\phi}(\alpha)}{1 - \beta\bar{\alpha}} = \int_\Gamma \frac{dv}{(1 - \bar{\alpha}e^{it})(1 - \beta e^{-it})}, \qquad \alpha, \beta \in D. \qquad (3\text{-}30)$$

We have dim $L^2(v) = $ *index of E.*

Proof of lemma. For any $z \in D$,

$$\operatorname{Re} \phi(z) = (2\pi)^{-1} \exp\left(\operatorname{Re} \pi i \int_E \frac{e^{it} + z}{e^{it} - z} \, d\sigma \right) \sin\left(\pi \int_E P(z, e^{it}) \, d\sigma \right).$$

Hence $\operatorname{Re} \phi(z) > 0$ on D. By Fatou's theorem, the sine factor tends nontangentially to $\sin(\pi \chi_E(e^{i\theta})) = 0$ σ-a.e., and hence $\operatorname{Re} \phi(e^{i\theta}) = 0$ σ-a.e. By the Riesz-Herglotz theorem, there is a nonnegative Borel measure v on Γ such that (3-30) holds. Since $\operatorname{Re} \phi(e^{i\theta}) = 0$ σ-a.e., v is singular by Fatou's theorem.

Suppose that the index of E is a positive integer n. Then modulo a σ-null set, $E = E_1 \cup \cdots \cup E_n$, where $E_j = \{e^{i\theta}: a_j \le \theta \le b_j\}$, $j = 1, \ldots, n$, and the arcs are proper and disjoint. By direct calculation,

$$\phi(z) = \frac{1}{2\pi i} e^{-i\pi\sigma(E)} \prod_1^n \frac{e^{ib_j} - z}{e^{ia_j} - z}.$$

Hence ϕ is rational with n simple poles, which all lie on Γ. Therefore v consists of n point masses, and so dim $L^2(v) = n = $ index of E.

Conversely, let dim $L^2(v)$ be a positive integer n. Then v consists of n point masses, and by (3-30), ϕ is a rational function with n simple poles, all on Γ. For $z \in D$,

$$\arg i\phi(z) = \pi \int_E P(z, e^{i\theta}) \, d\sigma,$$

where the argument is chosen in $[0, \pi]$. Passing to the boundary, we obtain

$$\arg i\phi(e^{i\theta}) = \pi \chi_E(e^{i\theta}) \qquad \sigma\text{-a.e.}$$

Since $\arg i\phi(e^{i\theta})$ is continuous except at the zeros and poles of ϕ, E is a union of intervals modulo a σ-null set. Hence by the argument of the preceding paragraph, dim $L^2(v) = n = $ index of E.

We omit the easy argument that E has index 0 if and only if $v = 0$. Once this is known, it follows that, in general, dim $L^2(v) = $ index of E. ∎

Proof of theorem. By nos. 4 and 6 above, T is absolutely continuous and sp$(T) = [c, d]$. Hence if T has spectral representation $T = \int x \, dE(x)$, then for any $f,g \in H^2(\Gamma)$,

$$\langle Tf,g \rangle_2 = \int_c^d x \frac{d}{dx} \langle E(x)f,g \rangle_2 \, dx. \tag{3-31}$$

The strategy of the proof is to use the generating formula for resolvents (3-24) to compute this for $f = k_\alpha$, $g = k_\beta$, $\alpha,\beta \in D$, where for any $\alpha \in D$,

$$k_\alpha(e^{i\theta}) = (1 - \bar{\alpha}e^{i\theta})^{-1} \quad \text{on } \Gamma.$$

Define a as in (3-25). By (3-24),

$$\frac{d}{dx} \langle E(x)k_\alpha,k_\beta \rangle_2 = \lim_{y\downarrow 0} \frac{y}{\pi} \int_{-\infty}^{\infty} \frac{d\langle E(t)k_\alpha,k_\beta \rangle_2}{(t-x)^2 + y^2}$$

$$= \lim_{y\downarrow 0} \frac{1}{2\pi i} \left(\langle (T-zI)^{-1}k_\alpha,k_\beta \rangle_2 - \langle (T-\bar{z}I)^{-1}k_\alpha,k_\beta \rangle_2 \right)$$

$$= \lim_{y\downarrow 0} \frac{1}{2\pi i} \frac{\bar{a}(\alpha,\bar{z})a(\beta,z) - \bar{a}(\alpha,z)a(\beta,\bar{z})}{1 - \bar{\alpha}\beta}$$

$$= \frac{1}{2\pi i} \frac{\bar{a}(\alpha,x - i0)a(\beta,x + i0) - \bar{a}(\alpha,x + i0)a(\beta,x - i0)}{1 - \bar{\alpha}\beta}$$

$$= \xi_\alpha(x)\bar{\xi}_\beta(x) \frac{\phi_x(\beta) + \bar{\phi}_x(\alpha)}{1 - \bar{\alpha}\beta}. \tag{3-32}$$

Here, for any $\alpha \in D$,

$$\xi_\alpha(x) = \bar{a}(\alpha,x + i0) \quad \text{and} \quad \phi_x(\alpha) = -\frac{1}{2\pi i} \frac{a(\alpha,x - i0)}{a(\alpha,x + i0)}.$$

The limits $a(\alpha,x \pm i0) = \lim_{y\downarrow 0} a(\alpha,x \pm iy)$ exist for all x satisfying (3-26):

$$a(\alpha,x \pm i0) = \exp\left(-\frac{1}{2} \int_\Gamma \frac{e^{it} + \alpha}{e^{it} - \alpha} \log|w(e^{it}) - x| \, d\sigma \right)$$

$$\cdot \exp\left(\pm \frac{1}{2}\pi i \int_{\Gamma \backslash E_x} \frac{e^{it} + \alpha}{e^{it} - \alpha} \, d\sigma \right),$$

where $E_x = \{e^{it}: w(e^{it}) \geq x\}$ as in (ii). Thus

$$\phi_x(\alpha) = -\frac{1}{2\pi i} \exp\left(-\pi i \int_{\Gamma \backslash E_x} \frac{e^{it} + \alpha}{e^{it} - \alpha} \, d\sigma \right)$$

$$= \frac{1}{2\pi i} \exp\left(\pi i \int_{E_x} \frac{e^{it} + \alpha}{e^{it} - \alpha} \, d\sigma \right).$$

By the lemma,

$$\frac{\phi_x(\beta) + \bar{\phi}_x(\alpha)}{1 - \bar{\alpha}\beta} = \int_\Gamma \frac{dv_x}{(1 - \bar{\alpha}e^{it})(1 - \beta e^{-it})} = \langle k_\alpha, k_\beta \rangle_{L^2(v_x)},$$

where v_x is a nonnegative singular Borel measure on Γ such that

$$\dim L^2(v_x) = \text{index of } E_x. \tag{3-33}$$

Hence (3-32) becomes

$$\frac{d}{dx} \langle E(x)k_\alpha, k_\beta \rangle_2 = \xi_\alpha(x)\bar{\xi}_\beta(x)\langle k_\alpha, k_\beta \rangle_{L^2(v_x)}, \tag{3-34}$$

and by (3-31),

$$\langle Tk_\alpha, k_\beta \rangle_2 = \int_c^d x\xi_\alpha(x)\bar{\xi}_\beta(x)\langle k_\alpha, k_\beta \rangle_{L^2(v_x)}\, dx. \tag{3-35}$$

Let $\mu = $ Lebesgue measure on $\text{sp}(T) = [c,d]$. Let D' be a countable set in D that has a limit point in D. For each x, the functions k_α, $\alpha \in D'$, span a dense set in $L^2(v_x)$ (since v_x is singular). Hence by the lemma in no. 7, there is a unique direct integral Hilbert space

$$\mathscr{K} = \int_{[c,d]} \oplus\, L^2(v_x)\, d\mu$$

that contains each function $x \to \xi_\alpha(x)k_\alpha$, $\alpha \in D'$. By the same lemma, $\mathscr{K}$ contains each function $x \to \xi_\alpha(x)k_\alpha$, $\alpha \in D$. By (3-34) there is a unique Hilbert space isomorphism $U: H^2(\Gamma) \to \mathscr{K}$ such that

$$U: k_\alpha \to \xi_\alpha(x)k_\alpha, \qquad \alpha \in D.$$

By (3-35), U establishes a unitary equivalence between T and multiplication by x on $\mathscr{K}$. In view of (3-33), the theorem follows.   ∎

9. *Diagonalization of self-adjoint Toeplitz operators that have multiplicity one.* We continue the discussion of no. 8, but we restrict w so that $T = T(w)$ has spectral multiplicity one. Thus we assume that there exist measurable functions $a(x)$, $b(x)$ such that, a.e. on $[c,d]$, $0 < b(x) - a(x) < 2\pi$ and

$$E_x = \{e^{it}: a(x) \le t \le b(x)\}$$

modulo a σ-null set.

We replace the direct integral space $\mathscr{K}$ of no. 8 by $L^2(\rho)$, where ρ is the measure on $[c,d]$ defined by

$$d\rho(x) = \pi^{-1}\sin((b(x) - a(x))/2)\, dx.$$

For almost all real x let $\psi(\cdot,x)$ be the unique outer function on D such that

$$|w(e^{it}) - x| = |\psi(e^{it},x)|^2 \qquad \sigma\text{-a.e.}$$

and $\psi(0,x) > 0$. Such a function exists for all x satisfying (3-26).

THEOREM.   *There is a unique isometry V mapping $H^2(\Gamma)$ onto $L^2(\rho)$ such that for all $\alpha \in D$,*

$$(Vk_\alpha)(x) = [\bar\psi(\alpha,x)(1 - \bar\alpha e^{ia(x)})^{\frac{1}{2}}(1 - \bar\alpha e^{ib(x)})^{\frac{1}{2}}]^{-1},$$

where $\{k_\alpha\}_{\alpha \in D}$ is as in no. 8. Furthermore:

   (i) *VTV^{-1} is multiplication by x on $L^2(\rho)$, and*
   (ii) *if $f \in L^2(\rho)$, then for all $\alpha \in D$,*

$$(V^{-1}f)(\alpha) = \int_c^d f(x)\,\overline{(Vk_\alpha)(x)}\,d\rho(x).$$

Proof.   This can be deduced from the constructions in no. 8 by showing that for all $\alpha,\beta \in D$,

$$\phi_x(\alpha) = \frac{1}{2\pi i}\,e^{\frac{1}{2}i(b - a)}\frac{1 - \alpha e^{-ib}}{1 - \alpha e^{-ia}},$$

and so

$$\frac{\phi_x(\beta) + \bar\phi_x(\alpha)}{1 - \bar\alpha\beta} = \rho'(x)(1 - \bar\alpha e^{ia})^{-1}(1 - \beta e^{-ia})^{-1}.$$

Then we obtain

$$\frac{d}{dx}\langle E(x)k_\alpha,k_\beta\rangle_2 = \rho'(x)[\bar\psi(\alpha,x)(1 - \bar\alpha e^{ia})^{\frac{1}{2}}(1 - \bar\alpha e^{ib})^{\frac{1}{2}}]^{-1}$$

$$\cdot [\psi(\beta,x)(1 - \beta e^{-ia})^{\frac{1}{2}}(1 - \beta e^{-ib})^{\frac{1}{2}}]^{-1},$$

and one can deduce the result from this. Alternatively, see Rosenblum [1962].   ∎

Example 1.   Let $w(e^{it}) = \cos t$. Then $[c,d] = [-1,1]$, and we can choose $b(x) = -a(x) = \arccos x$ for $-1 < x < 1$. Thus

$$\rho'(x) = \pi^{-1}(1 - x^2)^{\frac{1}{2}}, \qquad -1 < x < 1.$$

Since

$$|\cos t - x| = \tfrac{1}{2}|1 - 2xe^{it} + e^{2it}|,$$

we see by inspection that

$$\psi(\alpha,x) = 2^{-\frac{1}{2}}(1 - 2x\alpha + \alpha^2)^{\frac{1}{2}}.$$

It follows that

$$(Vk_\alpha)(x) = 2^{\frac{1}{2}}(1 - 2x\bar{\alpha} + \bar{\alpha}^2)^{-1}$$

$$= \sum_{n=0}^{\infty} 2^{\frac{1}{2}} U_n(x)\bar{\alpha}^n$$

for all $\alpha \in D$, where $\{U_n(x)\}_0^\infty$ are Chebychev polynomials (see Examples and Addenda to Chapter 1, no. 7). If $f \in L^2(\rho)$, then

$$(V^{-1}f)(\alpha) = \frac{2^{\frac{1}{2}}}{\pi} \int_{-1}^{1} f(x)(1 - 2x\alpha + \alpha^2)^{-1}(1 - x^2)^{\frac{1}{2}} \, dx,$$

$\alpha \in D$. The operator V diagonalizes $T(w)$.

The isomorphism theorem of no. 1 allows us to apply the diagonalization theory to any of the "natural" Toeplitz operators.

Example 2.   For fixed $k > 0$ consider the Wiener-Hopf operator

$$(T_k f)(x) = \int_0^{\infty} e^{-k|x-t|} f(t) \, dt, \qquad f \in L^2(0,\infty).$$

See Example (ii) in no. 1. Then T_k is diagonalized by the isometric operator U_k mapping $L^2(0,\infty)$ onto $L^2(v_k)$ such that

$$dv_k(\omega) = \frac{2}{\pi} \frac{d\omega}{\omega^2 + k^2} \quad \text{on } (0,\infty),$$

with

$$(U_k f)(\omega) = \int_0^{\infty} (\omega \cos \omega t + k \sin \omega t) f(t) \, dt$$

for each $f \in L^2(0,\infty) \cap L^1(0,\infty)$, and

$$(U_k^{-1} g)(t) = \int_0^{\infty} (\omega \cos \omega t + k \sin \omega t) g(\omega) \, dv_k(\omega)$$

for each $g \in L^2(v_k) \cap L^1(v_k)$. We find that

$$(U_k T_k U_k^{-1} g)(\omega) = \frac{2k}{\omega^2 + k^2} g(\omega)$$

for each $g \in L^2(v_k)$.

Remarks on the Literature.   The diagonalization of $T(2\cos\theta) = S + S^*$ was carried out by Hilbert [1912], p. 155, and Hellinger [1941], p. 158. Putnam [1959], [1967] gives an alternate approach to the spectral theory of Toeplitz operators using commutator theory. The spectral theory of unbounded Toeplitz operators is discussed in Clark [1967], Hartman [1963], Ismagilov [1963], Rosenblum [1965], and Rovnyak [1969]. Concerning a parallel theory for self-adjoint singular integral operators, see Rosenblum [1966], Pincus [1966], and Pincus and Rovnyak [1969]. The higher multiplicity case is more complicated and less complete; partial results are given in Pincus [1968] and Rovnyak [1972]. Some interesting results on the spectral theory of non-self-adjoint Toeplitz operators have been obtained by Clark [1980]. Rosenblum [1958I,II] has diagonalized the Hilbert matrix

$$H_k = [1/(m + n + 1 - k)]_{m,n=0}^{\infty}$$

for any real number k that is not a positive integer. This yields, in particular, an exact bound for the norm of H_k as an operator on l^2.

Notes

The abstract factorization theory for Toeplitz operators presented in Chapter 3 is from Rosenblum and Rovnyak [1971]. The solution of the abstract Szegö problem, Theorem 3.4, is due to Lowdenslager [1963] and Rosenblum [1968]. The comparison theorem is due to Moore [1969]. The abstract theory is based on concrete examples. For references to earlier literature, see the notes to Chapters 5 and 6. More general factorization problems are treated in Gellar and Page [1977], Moore, Rosenblum, and Rovnyak [1975], Page [1970], [1982], and Takahashi [1979].

Concerning the general theory of Toeplitz operators, see, for example, Brown and Halmos [1963], Douglas [1972], [1973], Grenander and Szegö [1958], Iohvidov [1982], and Widom [1965]. Additional references are given in these sources. The original work of Toeplitz on the subject appears in Toeplitz [1907], [1910], [1911a,b]. Otto Toeplitz has been called one of the founders of operator theory. Commemorative articles on his life and work may be found in Gohberg et al. [1981].

Chapter 4

Nevanlinna and Hardy Classes of Vector and Operator Valued Functions

4.1 Introduction

The purpose of this chapter is to set down the most basic facts concerning Nevanlinna and Hardy classes of vector and operator valued holomorphic functions. The emphasis in the chapter is on characterizations of the classes and boundary behavior (see Sections 4.2–4.6). After these ideas have been worked out, generalizations of many familiar results from the scalar theory follow in a routine way. Examples of such results are given in Sections 4.7 and 4.8. Great generality and completeness are not objectives of the chapter. In our choice of material we are guided mainly by what is needed for subsequent applications.

We assume familiarity with the scalar theory of Nevanlinna and Hardy classes on a disk or half-plane. What we need may be found, for example, in Duren [1970], Hoffman [1962], and Krylov [1939]. Other prerequisites from the theory of subharmonic functions are collected in the Appendix at the end of the book. Elementary properties of vector and operator valued functions are also taken for granted. These are given in Hille and Phillips [1957], Chapter III.

Throughout the chapter, X denotes a complex Banach space with norm $|\cdot|_X$. We write $\mathscr{C}$ for a separable Hilbert space and $|\cdot|_{\mathscr{C}}$, $\langle \cdot, \cdot \rangle_{\mathscr{C}}$ for the norm and inner product on $\mathscr{C}$. The norm on $\mathscr{B}(\mathscr{C})$, the space of bounded linear operators on $\mathscr{C}$, is written $|\cdot|_{\mathscr{B}(\mathscr{C})}$.

4.2 Nevanlinna and Hardy-Orlicz Classes

The key definitions are conveniently made in terms of harmonic majorants for subharmonic functions. To begin we show how subharmonic functions arise in the study of holomorphic functions with values in a Banach space.

THEOREM A.   *If $f(z)$ is a holomorphic X-valued function on a region $\Omega \subseteq \mathbf{C}$, then each of the functions listed below is subharmonic on Ω:*

 (i) $\log |f(z)|_X$,
 (ii) $\log^+ |f(z)|_X$,
 (iii) $|f(z)|_X^p$ *for $0 < p < \infty$, and*
 (iv) $\phi(\log^+ |f(z)|_X)$, *where ϕ is any nondecreasing convex function on $(-\infty, \infty)$.*

Here $\log^+ t = \max(\log t, 0)$ for $t > 0$ and $\log 0 = -\infty$.

Proof.   We show that $\log|f(z)|_X$ is subharmonic on Ω. Let $\bar{D}(a,r) \subseteq \Omega$, and let $p(z)$ be any polynomial such that $\log|f(z)|_X \le \operatorname{Re} p(z)$ on $\partial D(a,r)$. Then $|\exp(-p(z))f(z)|_X \le 1$ on $\partial D(a,r)$. By the maximum principle (Hille and Phillips [1957], p. 100), the same inequality extends to $D(a,r)$, and therefore $\log|f(z)|_X \le \operatorname{Re} p(z)$ on $D(a,r)$. Hence $\log|f(z)|_X$ is subharmonic on Ω (Appendix, Section 1).

Once it is known that $\log|f(z)|_X$ is subharmonic, it follows by standard properties of subharmonic functions that all of the functions listed in the theorem are subharmonic.   ∎

DEFINITION.   *Let $\Omega \subseteq \mathbf{C}$ be any region.*

   (i) *A holomorphic X-valued function $f(z)$ on Ω is of* bounded type *on Ω if $\log^+|f(z)|_X$ has a harmonic majorant on Ω. The class of all such functions is denoted $N_X(\Omega)$.*
   (ii) *If ϕ is any strongly convex function (Appendix, Section 3), then by $\mathscr{H}_{\phi,X}(\Omega)$ we mean the class of all holomorphic X-valued functions $f(z)$ on Ω such that $\phi(\log^+|f(z)|_X)$ has a harmonic majorant on Ω.*
   (iii) *We define $N_X^+(\Omega) = \bigcup \mathscr{H}_{\phi,X}(\Omega)$, where the union is over all strongly convex functions ϕ.*
   (iv) *By $H_X^\infty(\Omega)$ we mean the set of all bounded holomorphic X-valued functions on Ω.*

The sets $N_X(\Omega)$ and $N_X^+(\Omega)$ are called *Nevanlinna classes*, and $\mathscr{H}_{\phi,X}(\Omega)$ is a *Hardy-Orlicz* class. The term "bounded type" comes from the property expressed in Theorem D below.

When $X = \mathbf{C}$ in the absolute value norm, we drop the subscript X, and write simply $N(\Omega), N^+(\Omega), \mathscr{H}_\phi(\Omega), H^\infty(\Omega)$ for the above classes. We refer to this as the *scalar case.*

THEOREM B.   *For any region Ω and strongly convex function ϕ, $H_X^\infty(\Omega)$, $\mathscr{H}_{\phi,X}(\Omega)$, $N_X^+(\Omega)$, and $N_X(\Omega)$ are linear spaces and*

$$H_X^\infty(\Omega) \subseteq \mathscr{H}_{\phi,X}(\Omega) \subseteq N_X^+(\Omega) \subseteq N_X(\Omega). \tag{4-1}$$

Proof.   We use the elementary inequalities

$$\log^+(xy) \le \log^+ x + \log^+ y,$$

$$\log^+(x + y) \le \max(\log^+(2x), \log^+(2y)) \le \log^+ x + \log^+ y + \log 2$$

in the proof.

It is clear that $H_X^\infty(\Omega)$ is a linear space.

Let $f, g \in \mathscr{H}_{\phi,X}(\Omega)$, and let $\alpha \in \mathbf{C}$. Then

$$\phi(\log^+|\alpha f|_X) \le \phi(\log^+|f|_X + \log^+|\alpha|) \le M\phi(\log^+|f|_X) + K$$

for some constants $M \ge 0$ and $K \ge 0$ by properties of a strongly convex function (Appendix, Section 3). Since $f \in \mathscr{H}_{\phi,X}(\Omega)$, the right side has a harmonic majorant in Ω. Hence $\alpha f \in \mathscr{H}_{\phi,X}(\Omega)$. Examining separately the cases $|f(z)|_X \le |g(z)|_X$ and

$|f(z)|_X > |g(z)|_X$, we see that

$$\phi(\log^+|f + g|_X) \le \phi(\log^+|2f|_X) + \phi(\log^+|2g|_X).$$

It follows easily that $f + g \in \mathscr{H}_{\phi,X}(\Omega)$, and so $\mathscr{H}_{\phi,X}(\Omega)$ is a linear space.

The proof that $N_X^+(\Omega)$ is a linear space is straightforward once it is known that for any two strongly convex functions ψ_1 and ψ_2, there exists a strongly convex function ψ such that $\psi \le \psi_1$ and $\psi \le \psi_2$. To see this, recall that a convex function is the integral of a nondecreasing function. By the properties of a strongly convex function, we can write

$$\psi_j(x) = \int_{-\infty}^{x} g_j(t)\,dt + c_j, \qquad -\infty < x < \infty,$$

where g_j is nonnegative and nondecreasing on $(-\infty,\infty)$, $g_j(t) \to \infty$ as $t \to \infty$, and $c_j \ge 0$, $j = 1,2$. Construct in any way a nonnegative and nondecreasing function g on $(-\infty,\infty)$ such that $g(t) \to \infty$ as $t \to \infty$, $g \le g_1$ and $g \le g_2$ on $(-\infty,\infty)$, and $g(t + 1) \le 2g(t)$ for all real t. For example, there is a step function satisfying these conditions. Then in a straightforward way we see that

$$\psi(x) = \int_{-\infty}^{x} g(t)\,dt, \qquad -\infty < x < \infty,$$

has the required properties. It follows that $N_X^+(\Omega)$ is a linear space.

The fact that $N_X(\Omega)$ is a linear space follows from the inequalities

$$\log^+|\alpha f|_X \le \log^+|f|_X + \log^+|\alpha|,$$

$$\log^+|f + g|_X \le \log^+|f|_X + \log^+|g|_X + \log 2,$$

which hold for any $f, g \in N_X(\Omega)$ and $\alpha \in \mathbf{C}$.

The first two inclusions in (4-1) are obvious. If $f \in N_X^+(\Omega)$, then $f \in \mathscr{H}_{\psi,X}(\Omega)$ for some strongly convex function ψ. Choose $a > 0$ such that $\psi(t)/t \ge 1$ for $t > a$. Then $\log^+|f|_X \le \psi(\log^+|f|_X) + a$. Therefore $f \in N_X(\Omega)$, and the third inclusion of (4-1) follows. ∎

THEOREM C. *Let f belong to one of the classes $H_X^\infty(\Omega)$, $\mathscr{H}_{\phi,X}(\Omega)$, $N_X^+(\Omega)$, or $N_X(\Omega)$ for some region Ω.*

   (i) *If h is holomorphic on a region Ω' and $h(\Omega') \subseteq \Omega$, then $f \circ h$ belongs to the corresponding class on Ω'.*
   (ii) *If Ω'' is a region contained in Ω, then $f|\Omega''$ belongs to the corresponding class on Ω''.*

Proof. The assertions are immediate from the definitions of the classes. ∎

THEOREM D. *Let f be a holomorphic X-valued function on a region Ω.*

   (i) *A sufficient condition for f to belong to $N_X(\Omega)$ is that $f = g/u$, where $g \in H_X^\infty(\Omega)$ and u is a scalar valued holomorphic function satisfying $0 < |u| \le 1$ on Ω.*
   (ii) *If Ω is simply connected, then the sufficient condition of (i) is also necessary.*

Proof.   (i) Let $f = g/u$ as in (i). We can assume without loss of generality that $|g|_X \leq 1$ on Ω. Then

$$\log^+|f|_X \leq \log^+|g|_X + \log^+(1/|u|) = -\log|u|$$

on Ω. Since $-\log|u|$ is harmonic, $f \in N_X(\Omega)$. This proves (i).

(ii) Assume that Ω is simply connected and $f \in N_X(\Omega)$. Let h be a harmonic majorant for $\log^+|f|_X$. For each disk $D(a,r) \subseteq \Omega$ there is a holomorphic function $k_{a,r}$ on $D(a,r)$ such that $\operatorname{Re} k_{a,r} = h$ on $D(a,r)$. By the monodromy theorem (Rudin [1966], p. 319), there is a holomorphic function k on Ω such that $\operatorname{Re} k = h$ on Ω. Then $f = g/u$, where $g = fe^{-k}$ and $u = e^{-k}$ have the required properties. For example,

$$\log|f|_X \leq \log^+|f|_X \leq h = \operatorname{Re} k$$

and so $|g|_X = |fe^{-k}|_X \leq 1$. The theorem follows.   ∎

4.3   The Cases $\Omega = D$ and Π; Characterizations of the Classes

When the region Ω in Section 4.2 is a disk or half-plane, the defining properties for the Nevanlinna and Hardy-Orlicz classes have useful equivalent forms.

THEOREM A.   *Let $\Omega = D$ or Π, and let f be a holomorphic X-valued function on Ω. The following are equivalent*:

(i) *f is of bounded type, that is, $f \in N_X(\Omega)$;*
(ii) *$\log^+|f|_X$ has a harmonic majorant on Ω;*
(iii) *according as $\Omega = D$ or Π,*

$$\sup_{0<r<1} \int_\Gamma \log^+|f(re^{i\theta})|_X \, d\sigma < \infty \tag{4-2}$$

or

$$\sup_{y>0} \int_{-\infty}^\infty \frac{\log^+|f(x+iy)|_X}{x^2+(y+1)^2} \, dx < \infty; \tag{4-3}$$

(iv) *$f = g/u$, where g is a bounded holomorphic X-valued function on Ω and u is a scalar valued holomorphic function such that $0 < |u| \leq 1$ on Ω.*

Proof.   The equivalence of (i) and (ii) is by the definition of the classes $N_X(\Omega)$ in Section 4.2. Concerning (iii) see the criteria for the existence of harmonic majorants in the Appendix (Sections 1 and 4), and for (iv) use Theorem D in Section 4.2.   ∎

THEOREM B.   *Let $\Omega = D$ or Π, and let ϕ be a strongly convex function. If f is a holomorphic X-valued function on Ω, then the following are equivalent*:

(i) *$f \in \mathcal{H}_{\phi,X}(\Omega)$;*
(ii) *$\phi(\log^+|f|_X)$ has a harmonic majorant on Ω;*

(iii) *according as* $\Omega = D$ *or* Π,

$$\sup_{0 < r < 1} \int_{\Gamma} \phi(\log^+ |f(re^{i\theta})|_X)\, d\sigma < \infty \tag{4-4}$$

or

$$\sup_{y > 0} \int_{-\infty}^{\infty} \frac{\phi(\log^+ |f(x + iy)|_X)}{x^2 + (y + 1)^2}\, dx < \infty; \tag{4-5}$$

(iv) *same as* (iii), *but with* "$\log^+$" *replaced by* "$\log$".

Proof. The equivalence of (i), (ii), and (iii) follows from the definition in Section 4.2 and the criteria for the existence of harmonic majorants in the Appendix (Sections 1 and 4). Since

$$\phi(\log|f|_X) \le \phi(\log^+ |f|_X) \le \phi(\log|f|_X) + \phi(0),$$

(iii) is equivalent to (iv). ∎

THEOREM C. *Let* $\Omega = D$ *or* Π, *and let* f *be a holomorphic* X-*valued function on* Ω. *The following are equivalent:*

(i) $f \in N_X^+(\Omega)$;

(ii) $f \in \mathcal{H}_{\phi, X}(\Omega)$ *for some strongly convex function* ϕ;

(iii) $f = h/v$, *where* h *is a bounded holomorphic* X-*valued function on* Ω *and* v *is a scalar valued outer function such that* $0 < |v| \le 1$ *on* Ω.

Moreover, in the case $\Omega = D$, (i)–(iii) *are equivalent to:*

(iv) *the functions* $\{\log^+ |f(re^{i\theta})|_X\}_{0 < r < 1}$ *are uniformly integrable with respect to normalized Lebesgue measure* σ *on* Γ.

See the Appendix, Section 3, for the definition of a uniformly integrable family of functions.

Proof. The equivalence of (i) and (ii) is by the definition in Section 4.2. When $\Omega = D$, the equivalence of (ii) and (iv) follows from a theorem of de la Vallée Poussin and Nagumo (Appendix, Section 3). It remains to show that (ii) and (iii) are equivalent. It is sufficient to treat the case $\Omega = D$ since the other case then follows by conformal mapping (see Theorem C(i) of Section 4.2).

Assume (iii). Without loss of generality we can further assume that $|h|_X \le 1$ on D. Since v is an outer function,

$$\log^+ |f(z)|_X \le \log^+ |h(z)|_X + \log^+(1/|v(z)|)$$

$$= -\log|v(z)| = -\int_{\Gamma} P(z, e^{it}) \log|v(e^{it})|\, d\sigma$$

on D. The family consisting of the single function $-\log|v(e^{it})|$ in $L^1(\sigma)$ is uniformly integrable, so by the theorem of de la Vallée Poussin and Nagumo

(Appendix, Section 3), there is a strongly convex function ϕ such that

$$\int_\Gamma \phi(-\log|v(e^{it})|)\, d\sigma < \infty.$$

By Jensen's inequality (Rudin [1966], p. 61),

$$\phi(\log^+|f(z)|_X) \leq \phi\left(-\int_\Gamma P(z,e^{it})\log|v(e^{it})|\, d\sigma\right)$$

$$\leq \int_\Gamma P(z,e^{it})\phi(-\log|v(e^{it})|)\, d\sigma,$$

and hence

$$\int_\Gamma \phi(\log^+|f(re^{i\theta})|_X)\, d\sigma(e^{i\theta})$$

$$\leq \int_\Gamma\int_\Gamma P(re^{i\theta},e^{it})\, d\sigma(e^{i\theta})\phi(-\log|v(e^{it})|)\, d\sigma(e^{it})$$

$$= \int_\Gamma \phi(-\log|v(e^{it})|)\, d\sigma(e^{it}).$$

It follows that $\phi(\log^+|f(z)|_X)$ has a harmonic majorant (Appendix, Section 1). Therefore $f \in \mathscr{H}_{\phi,X}(D)$ and (ii) holds.

Assume (ii) and exclude the trivial case $f \equiv 0$. It is easy to see that the function $u(z) = |f(z)|_X$ satisfies the hypotheses of the Szegö-Solomentsev theorem (Appendix, Section 2). The inequality in the second part of that theorem may be written

$$|f|_X \leq |gS_+/S_-|,$$

where g is a scalar valued outer function and S_+, S_- are scalar valued singular inner functions. By our assumption (ii) and the third part of the Szegö-Solomentsev theorem, S_+ is a constant of modulus 1. Choose an outer function v such that $0 < |v| \leq 1$ and $|vg| \leq 1$ on D. Setting $h = vf$, we obtain $f = h/v$ as required in (iii). ∎

4.4 Hardy Classes

We define $H^p_X(D) = \mathscr{H}_{\phi,X}(D)$, where $\phi(t) = e^{pt}$, $0 < p < \infty$. The class $H^\infty_X(D)$ has previously been defined in Section 4.2.

THEOREM A.   *Let $0 < p < \infty$. If f is a holomorphic X-valued function on D, then the following are equivalent:*

(i) *$f \in H^p_X(D)$;*
(ii) *$|f|^p_X$ has a harmonic majorant on D;*
(iii) *$\sup_{0<r<1} \int_\Gamma |f(re^{i\theta})|^p_X\, d\sigma < \infty$.*

Proof. By Section 4.2, Theorem A, $|f|_X^p$ is subharmonic on D. It is easy to see that $|f|_X^p$ has a harmonic majorant on D if and only if $\exp(p\log^+|f|_X)$ has a harmonic majorant on D. Hence the result follows from the definition of $H_X^p(D)$ and the condition for the existence of a harmonic majorant for a subharmonic function on D (Appendix, Section 1). ∎

For each $f \in H_X^p(D)$, $0 < p < \infty$, set

$$\|f\|_p = \sup_{0 < r < 1}\left(\int_\Gamma |f(re^{i\theta})|_X^p \, d\sigma\right)^{1/p}.$$

Since $|f(z)|_X^p$ is subharmonic on D, we can also write this as

$$\|f\|_p = \lim_{r \uparrow 1}\left(\int_\Gamma |f(re^{i\theta})|_X^p \, d\sigma\right)^{1/p}.$$

For $f \in H_X^\infty(D)$ set

$$\|f\|_\infty = \sup_{z \in D}|f(z)|_X.$$

As in the scalar case (Duren [1970], Chapter 11), we define two kinds of Hardy classes on the upper half-plane Π.

First kind. Let $H_X^p(\Pi)$, $0 < p < \infty$, be the set of all holomorphic X-valued functions F on Π such that

$$\|F\|_p = \sup_{y > 0}\left(\int_{-\infty}^\infty |F(x + iy)|_X^p \, dx\right)^{1/p} < \infty.$$

The class $H_X^\infty(\Pi)$ is as previously defined in Section 4.2. For any $F \in H_X^\infty(\Pi)$ set

$$\|F\|_\infty = \sup_{y > 0}|F(z)|_X.$$

Second kind. Let $\mathscr{H}_X^p(\Pi)$, $0 < p \le \infty$, be the set of all holomorphic X-valued functions F on Π such that $F \circ \alpha \in H_X^p(D)$, where α is the mapping of D on Π given by

$$\alpha: w \to i(1 + w)/(1 - w).$$

For each $F \in \mathscr{H}_X^p(\Pi)$ set

$$\|\|F\|\|_p = \|F \circ \alpha\|_p.$$

THEOREM B. *Let* $0 < p < \infty$. *If* F *is a holomorphic* X-*valued function on* Π, *then the following are equivalent:*

(i) $F \in \mathscr{H}_X^p(\Pi)$;

(ii) $|F|_X^p$ *has a harmonic majorant on* Π;

(iii) $\displaystyle\sup_{y > 0}\int_{-\infty}^\infty \frac{|F(x + iy)|_X^p}{x^2 + (y + 1)^2}\,dx < \infty.$

Proof.   The equivalence of (i) and (ii) follows from Theorem A and the definition of $\mathscr{H}_X^{\,p}(\Pi)$. The equivalence of (ii) and (iii) is by the theorem of Flett and Kuran in the Appendix, Section 4.   ∎

THEOREM C.   *If* $0 < p < \infty$, *then*

$$H_X^{\infty}(D) \subseteq H_X^{p}(D) \subseteq N_X^{+}(D) \qquad and \qquad H_X^{p}(\Pi) \subseteq \mathscr{H}_X^{\,p}(\Pi) \subseteq N_X^{+}(\Pi).$$

Proof.   The inclusions follow easily from the definitions and characterizations of the classes given above.   ∎

4.5   The Cases $X = \mathscr{C}$ and $\mathscr{B}(\mathscr{C})$; Fatou's Theorem for Bounded Functions

From now on we assume that X is either the Hilbert space $\mathscr{C}$ or the space $\mathscr{B}(\mathscr{C})$ of bounded linear operators on $\mathscr{C}$.

In this section we prove a version of Fatou's theorem on the existence of nontangential limits at boundary points. Boundary behavior in general is discussed in the next section.

FATOU'S THEOREM.   (i) *Each* f *in* $H_{\mathscr{C}}^{\infty}(D)$ *has a nontangential limit*

$$f(e^{i\theta}) = \lim_{z \to e^{i\theta}} f(z)$$

σ-*a.e. in the strong topology of* $\mathscr{C}$.

(ii) *Each* F *in* $H_{\mathscr{B}(\mathscr{C})}^{\infty}(D)$ *has a nontangential limit*

$$F(e^{i\theta}) = \lim_{z \to e^{i\theta}} F(z)$$

σ-*a.e. in the strong operator topology on* $\mathscr{B}(\mathscr{C})$.

We digress briefly to review some notions from measure theory. As this material is widely used and generally well known, we will merely give definitions and state the facts we need without proof.

Let $(A,\mathscr{F},\mu)$ be a measure space. A $\mathscr{C}$-valued function f on A is *weakly measurable* if $\langle f(\cdot),a \rangle_{\mathscr{C}}$ is measurable for each $a \in \mathscr{C}$. A $\mathscr{B}(\mathscr{C})$-valued function F on A is *weakly measurable* if $\langle F(\cdot)a,b \rangle_{\mathscr{C}}$ is measurable for all $a,b \in \mathscr{C}$.

Let f,g be weakly measurable $\mathscr{C}$-valued functions, and let F,G be weakly measurable $\mathscr{B}(\mathscr{C})$-valued functions on A. Then Ff is a weakly measurable $\mathscr{C}$-valued function, FG is a weakly measurable $\mathscr{B}(\mathscr{C})$-valued function, and $\langle f(\cdot),g(\cdot) \rangle_{\mathscr{C}}$, $|f(\cdot)|_{\mathscr{C}}$, and $|F(\cdot)|_{\mathscr{B}(\mathscr{C})}$ are measurable scalar valued functions on A.

Let f be a weakly measurable $\mathscr{C}$-valued function and F a weakly measurable $\mathscr{B}(\mathscr{C})$-valued function on A such that

$$\int_A |f|_{\mathscr{C}} \, d\mu < \infty \qquad and \qquad \int_A |F|_{\mathscr{B}(\mathscr{C})} \, d\mu < \infty.$$

We define

$$\int_A f \, d\mu = c \qquad and \qquad \int_A F \, d\mu = C,$$

where $c \in \mathscr{C}$ is the unique vector and $C \in \mathscr{B}(\mathscr{C})$ is the unique operator such that

$$\int_A \langle f(t),a \rangle_\mathscr{C} \, d\mu = \langle c,a \rangle_\mathscr{C} \qquad \text{and} \qquad \int_A \langle F(t)a,b \rangle_\mathscr{C} \, d\mu = \langle Ca,b \rangle_\mathscr{C}$$

for all $a,b \in \mathscr{C}$. Then

$$\left| \int_A f \, d\mu \right|_\mathscr{C} \leq \int_A |f|_\mathscr{C} \, d\mu \qquad \text{and} \qquad \left| \int_A F \, d\mu \right|_{\mathscr{B}(\mathscr{C})} \leq \int_A |F|_{\mathscr{B}(\mathscr{C})} \, d\mu.$$

Integrals defined in this way are said to converge in the *weak sense*.

Let $X = \mathscr{C}$ or $\mathscr{B}(\mathscr{C})$. Define $L^p_X(\mu)$, $0 < p < \infty$, as the space of weakly measurable X-valued functions f on A such that

$$\|f\|_p = \left(\int_A |f|^p_X \, d\mu \right)^{1/p} < \infty.$$

Let $L^\infty_X(\mu)$ be the space of weakly measurable X-valued functions f on A with

$$\|f\|_\infty = \operatorname*{ess\,sup}_A |f|_X < \infty.$$

For any p, $1 \leq p \leq \infty$, $L^p_X(\mu)$ is a Banach space, and $L^2_\mathscr{C}(\mu)$ is a Hilbert space in the inner product

$$\langle f,g \rangle_2 = \int_A \langle f(t),g(t) \rangle_\mathscr{C} \, d\mu,$$

$f,g \in L^2_\mathscr{C}(\mu)$.

We follow the standard conventions of measure theory. For example, a function that is defined almost everywhere on a set is regarded as defined on the set, and two functions that are equal almost everywhere are usually identified. For a full treatment of measure and integration theory for vector and operator valued functions, including strong as well as weak notions, see Hille and Phillips [1957], Chapter III.

Proof of Fatou's theorem. The proof in effect reduces the vector and operator version of the theorem to the scalar version (Duren [1970], Hoffman [1962]).

(i) Let $f \in H^\infty_\mathscr{C}(D)$. Then $f \in H^2_\mathscr{C}(D)$ and by Section 1.15 the coefficients in the Taylor expansion $f(z) = \sum_0^\infty a_j z^j$ satisfy $\sum_0^\infty |a_j|^2_\mathscr{C} < \infty$. Therefore we may define $\phi \in L^2_\mathscr{C}(\sigma)$ by

$$\phi(e^{it}) = \sum_0^\infty a_j e^{ijt},$$

where the series converges in the metric of $L^2_\mathscr{C}(\sigma)$. For any $c \in \mathscr{C}$ and $z \in D$,

$$\langle f(z),c \rangle_\mathscr{C} = \int_\Gamma P(z,e^{it}) \langle \phi(e^{it}),c \rangle_\mathscr{C} \, d\sigma$$

and so

$$f(z) = \int_{\Gamma} P(z,e^{it})\phi(e^{it})\,d\sigma. \tag{4-6}$$

Let $\mathscr{E}$ be a countable dense set in $\mathscr{C}$. By the scalar version of Fatou's theorem, there is a σ-null set $N \subseteq \Gamma$ such that

$$\lim_{z \to e^{i\theta}} \int_{\Gamma} P(z,e^{it})|\phi(e^{it}) - a|_{\mathscr{C}}\,d\sigma = |\phi(e^{i\theta}) - a|_{\mathscr{C}} \tag{4-7}$$

nontangentially for each $e^{i\theta} \in \Gamma\backslash N$ and $a \in \mathscr{E}$.

Fix $e^{i\theta} \in \Gamma\backslash N$. Let S be an open triangular sector in D with vertex $e^{i\theta}$. Given $\varepsilon > 0$, choose $a \in \mathscr{E}$ such that $|\phi(e^{i\theta}) - a|_{\mathscr{C}} < \varepsilon/2$. By (4-6),

$$|f(z) - \phi(e^{i\theta})|_{\mathscr{C}} = \left| \int_{\Gamma} P(z,e^{it})[\phi(e^{it}) - a + a - \phi(e^{i\theta})]\,d\sigma \right|_{\mathscr{C}}$$

$$\leq \int_{\Gamma} P(z,e^{it})|\phi(e^{it}) - a|_{\mathscr{C}}\,d\sigma + |\phi(e^{i\theta}) - a|_{\mathscr{C}}.$$

Hence by (4-7),

$$\limsup_{\substack{z \to e^{i\theta} \\ z \in S}} |f(z) - \phi(e^{i\theta})|_{\mathscr{C}} \leq 2|\phi(e^{i\theta}) - a|_{\mathscr{C}} < \varepsilon.$$

By the arbitrariness of ε, $|f(z) - \phi(e^{i\theta})|_{\mathscr{C}} \to 0$ for $z \in S$, $z \to e^{i\theta}$.

We thus obtain (i) with $f(e^{i\theta}) = \phi(e^{i\theta})$ for all $e^{i\theta} \in \Gamma\backslash N$.

(ii) Let $\mathscr{E}$ be as above, and apply (i) to $F(z)a$ for each fixed $a \in \mathscr{E}$. Since $\mathscr{E}$ is countable, there is a σ-null set $N \subseteq \Gamma$ such that

$$\lim_{z \to e^{i\theta}} F(z)a = \phi_a(e^{i\theta})$$

exists nontangentially for all $e^{i\theta} \in \Gamma\backslash N$ and $a \in \mathscr{E}$.

Fix $e^{i\theta} \in \Gamma\backslash N$. Define $s_0(e^{i\theta};\cdot,\cdot)$ on $\mathscr{E} \times \mathscr{E}$ by

$$s_0(e^{i\theta};a,b) = \langle \phi_a(e^{i\theta}),b \rangle_{\mathscr{C}}, \qquad a,b \in \mathscr{E}.$$

For any $a_1,a_2,b_1,b_2 \in \mathscr{E}$,

$$|s_0(e^{i\theta};a_1,b_1) - s_0(e^{i\theta};a_2,b_2)|$$

$$= \lim_{z \to e^{i\theta}} |\langle F(z)a_1,b_1 - b_2 \rangle_{\mathscr{C}} + \langle F(z)(a_1 - a_2),b_2 \rangle_{\mathscr{C}}|$$

$$\leq \|F\|_{\infty}|a_1|_{\mathscr{C}}|b_1 - b_2|_{\mathscr{C}} + \|F\|_{\infty}|a_1 - a_2|_{\mathscr{C}}|b_2|_{\mathscr{C}}.$$

Therefore $s_0(e^{i\theta};\cdot,\cdot)$ has a unique extension by continuity $s(e^{i\theta};\cdot,\cdot)$ to $\mathscr{C} \times \mathscr{C}$. By construction,

$$s(e^{i\theta};a,b) = \lim_{z \to e^{i\theta}} \langle F(z)a,b \rangle_{\mathscr{C}}$$

nontangentially for all $a,b \in \mathcal{E}$. Routine arguments now show that $s(e^{i\theta};\cdot,\cdot)$ is a bounded sesquilinear form on $\mathcal{C}$ with $\|s\| \leq \|F\|_\infty$. Hence there is an operator $F(e^{i\theta}) \in \mathcal{B}(\mathcal{C})$ such that $|F(e^{i\theta})|_{\mathcal{B}(\mathcal{C})} \leq \|F\|_\infty$ and

$$s(e^{i\theta};a,b) = \langle F(e^{i\theta})a,b \rangle_{\mathcal{C}}, \qquad a,b \in \mathcal{C}.$$

It follows that $F(z)a \to F(e^{i\theta})a$ nontangentially in the norm of $\mathcal{C}$ for all $a \in \mathcal{E}$. Since $F(z)$ is bounded on D, the same holds for all $a \in \mathcal{C}$. The result follows. ∎

4.6   Boundary Behavior of Functions of Bounded Type

The boundary properties of vector and operator valued functions of bounded type are very similar to the scalar theory. The most serious loss is that in the case of operator valued functions, Fatou's theorem fails in the norm topology (Examples and Addenda, no. 1). However, as we have seen in Section 4.5, Fatou's theorem holds relative to the strong operator topology, and this is an adequate substitute.

THEOREM A.   *Let $X = \mathcal{C}$ or $\mathcal{B}(\mathcal{C})$. For each $f \in N_X(D)$ a nontangential limit*

$$f(e^{i\theta}) = \lim_{z \to e^{i\theta}} f(z) \tag{4-8}$$

exists σ-a.e. on Γ in the strong topology if $X = \mathcal{C}$ and strong operator topology if $X = \mathcal{B}(\mathcal{C})$. Also, for $X = \mathcal{C}$ or $\mathcal{B}(\mathcal{C})$,

$$|f(e^{i\theta})|_X = \lim_{z \to e^{i\theta}} |f(z)|_X \tag{4-9}$$

nontangentially σ-a.e. on Γ. Moreover, if $f \not\equiv 0$ on D, then

$$\log|f(e^{i\theta})|_X \in L^1(\sigma). \tag{4-10}$$

Proof.   Assume $f \not\equiv 0$. By Section 4.3, Theorem A(iv), we may further assume that f is bounded on D. Then the existence of a nontangential boundary function (4-8) follows from Fatou's theorem in Section 4.5. We obtain (4-10) from the Szegö-Solomentsev theorem applied to the function $u(z) = |f(z)|_X$ (Appendix, Section 2).

When $X = \mathcal{C}$, (4-9) is clear. It remains to prove (4-9) when $X = \mathcal{B}(\mathcal{C})$. Choose a σ-null set $N \subseteq \Gamma$ such that for each $e^{i\theta} \in \Gamma \backslash N$, $f(e^{i\theta})$ exists and is nonzero and

$$\lim_{z \to e^{i\theta}} \int_\Gamma P(z,e^{it}) \log|f(e^{it})|_{\mathcal{B}(\mathcal{C})}\, d\sigma = \log|f(e^{i\theta})|_{\mathcal{B}(\mathcal{C})} \tag{4-11}$$

nontangentially. Fix $e^{i\theta} \in \Gamma \backslash N$, and let S be a triangular sector in D with vertex at $e^{i\theta}$. For all $z \in D$,

$$\log|f(z)|_X \leq \int_\Gamma P(z,e^{it}) \log|f(e^{it})|_X\, d\sigma$$

(Examples and Addenda, no. 3; we still assume that f is bounded and so this result is applicable). Hence by (4-11),

$$\limsup_{\substack{z \to e^{i\theta} \\ z \in S}} |f(z)|_{\mathscr{B}(\mathscr{C})} \leq |f(e^{i\theta})|_{\mathscr{B}(\mathscr{C})}.$$

For any $a,b \in \mathscr{C}$, $|a|_{\mathscr{C}} = |b|_{\mathscr{C}} = 1$,

$$\langle f(e^{i\theta})a,b \rangle_{\mathscr{C}} = \lim_{\substack{z \to e^{i\theta} \\ z \in S}} |\langle f(z)a,b \rangle_{\mathscr{C}}| \leq \liminf_{\substack{z \to e^{i\theta} \\ z \in S}} |f(z)|_{\mathscr{B}(\mathscr{C})}.$$

By the arbitrariness of a and b,

$$|f(e^{i\theta})|_{\mathscr{B}(\mathscr{C})} \leq \liminf_{\substack{z \to e^{i\theta} \\ z \in S}} |f(z)|_{\mathscr{B}(\mathscr{C})}.$$

Therefore,

$$|f(e^{i\theta})|_{\mathscr{B}(\mathscr{C})} = \lim_{\substack{z \to e^{i\theta} \\ z \in S}} |f(z)|_{\mathscr{B}(\mathscr{C})}.$$

Since the sector S is arbitrary, the result follows. ∎

THEOREM B. *Let $X = \mathscr{C}$ or $\mathscr{B}(\mathscr{C})$. For each $f \in N_X(\Pi)$, a nontangential limit*

$$f(x) = \lim_{z \to x} f(z) \tag{4-12}$$

exists a.e. on $(-\infty,\infty)$ in the strong topology if $X = \mathscr{C}$ and strong operator topology if $X = \mathscr{B}(\mathscr{C})$. Also, for $X = \mathscr{C}$ or $\mathscr{B}(\mathscr{C})$,

$$|f(x)|_X = \lim_{z \to x} |f(z)|_X \tag{4-13}$$

nontangentially a.e. on $(-\infty,\infty)$. Moreover, either $f \equiv 0$ on Π or

$$\int_{-\infty}^{\infty} \frac{|\log|f(t)|_X|}{1+t^2} \, dt < \infty. \tag{4-14}$$

Proof. The mapping $\alpha: w \to i(1 + w)/(1 - w)$ takes D onto Π and $\Gamma \backslash \{1\}$ onto $(-\infty,\infty)$. It is conformal at all points of $\bar{D} \backslash \{1\}$. We thus obtain Theorem B from Theorem A by a routine change of variables. ∎

THEOREM C. *Let $X = \mathscr{C}$ or $\mathscr{B}(\mathscr{C})$, and let ϕ be a strongly convex function.*
(i) *If $f \in N_X^+(D)$, then $f \in \mathscr{H}_{\phi,X}(D)$ if and only if*

$$\int_{\Gamma} \phi(\log|f(e^{it})|_X) \, d\sigma < \infty. \tag{4-15}$$

In this case,

$$\lim_{r \uparrow 1} \int_{\Gamma} \phi(\log|f(re^{it})|_X) \, d\sigma = \int_{\Gamma} \phi(\log|f(e^{it})|_X) \, d\sigma. \tag{4-16}$$

(ii) *If $X = \mathscr{C}$ and $\lim_{t \to -\infty} \phi(t) = 0$, then also for each $f \in \mathscr{H}_{\phi,\mathscr{C}}(D)$,*

$$\lim_{r \uparrow 1} \int_{\Gamma} \phi(\log|f(e^{i\theta}) - f(re^{i\theta})|_{\mathscr{C}}) \, d\sigma = 0. \tag{4-17}$$

Assertion (ii) fails if $\mathscr{C}$ is replaced by $\mathscr{B}(\mathscr{C})$ (Examples and Addenda, no. 1).

Proof. (i) Let $f \in N_X^+(D)$ and assume $f \not\equiv 0$. If $f \in \mathscr{H}_{\phi,X}(D)$, then by Section 4.3, Theorem B, and Fatou's lemma,

$$\int_{\Gamma} \phi(\log|f(e^{it})|_X) \, d\sigma \leq \lim_{r \uparrow 1} \int_{\Gamma} \phi(\log|f(re^{it})|_X) \, d\sigma$$

$$\leq \lim_{r \uparrow 1} \int_{\Gamma} \phi(\log^+|f(re^{it})|_X) \, d\sigma < \infty. \tag{4-18}$$

Thus (4-15) holds.

Conversely suppose that (4-15) holds. For all $z \in D$,

$$\log|f(z)|_X \leq \int_{\Gamma} P(z,e^{it}) \log|f(e^{it})|_X \, d\sigma$$

(Examples and Addenda, no. 3). By Jensen's inequality (Rudin [1966], p. 61),

$$\phi(\log|f(z)|_X) \leq \int_{\Gamma} P(z,e^{it})\phi(\log|f(e^{it})|_X) \, d\sigma.$$

Hence for $0 < r < 1$,

$$\int_{\Gamma} \phi(\log|f(re^{i\theta})|_X) \, d\sigma(e^{i\theta})$$

$$\leq \int_{\Gamma}\int_{\Gamma} P(re^{i\theta},e^{it}) \, d\sigma(e^{i\theta})\phi(\log|f(e^{it})|_X) \, d\sigma(e^{it}) = \int_{\Gamma} \phi(\log|f(e^{it})|_X) \, d\sigma(e^{it}). \tag{4-19}$$

A similar argument with "log" replaced by "log$^+$" yields

$$\int_{\Gamma} \phi(\log^+|f(re^{i\theta})|_X) \, d\sigma \leq \int_{\Gamma} \phi(\log^+|f(e^{it})|_X) \, d\sigma,$$

where by (4-15) the integral on the right is finite. In particular, $f \in \mathscr{H}_{\phi,X}(D)$ by Section 4.3, Theorem B. Moreover, we obtain (4-16) by combining (4-18) and (4-19).

(ii) Let $f \in \mathscr{H}_{\phi,\mathscr{C}}(D)$, $f \not\equiv 0$, and assume that $\phi(-\infty) = 0$, where $\phi(-\infty) = \lim_{x \to -\infty} \phi(x)$. Fix a sequence $r_n \uparrow 1$ and set $f_n(e^{i\theta}) = f(r_n e^{i\theta})$ on Γ for each

$n = 1,2,3,\ldots$. Let $\varepsilon > 0$ be given. By (4-15) there is a $\delta > 0$ such that $\delta < \varepsilon$ and

$$\int_\Delta \phi(\log|f|_\mathscr{C})\,d\sigma < \varepsilon \tag{4-20}$$

for every Borel set $\Delta \subseteq \Gamma$ with $\sigma(\Delta) < \delta$. By Egoroff's theorem we can choose Δ such that $\sigma(\Delta) < \delta$ and $|f - f_n|_\mathscr{C} \to 0$ uniformly on $\Gamma\backslash\Delta$. Since $\phi(-\infty) = 0$, $\phi(\log|f - f_n|_\mathscr{C}) \to 0$ uniformly on $\Gamma\backslash\Delta$, and so

$$\lim_{n\to\infty} \int_{\Gamma\backslash\Delta} \phi(\log|f - f_n|_\mathscr{C})\,d\sigma = 0. \tag{4-21}$$

By the definition of a strongly convex function (Appendix, Section 3, fifth property), there exist constants $M \geq 0$ and $K \geq 0$ such that $\phi(t + \log 2) \leq M\phi(t) + K$ for all real t. Then for all $n = 1,2,3,\ldots$,

$$\phi(\log|f - f_n|_\mathscr{C}) = \phi(\log(\tfrac{1}{2}|f - f_n|_\mathscr{C}) + \log 2)$$
$$\leq M\phi(\log(\tfrac{1}{2}|f - f_n|_\mathscr{C})) + K$$
$$\leq M\max(\phi(\log|f|_\mathscr{C}),\phi(\log|f_n|_\mathscr{C})) + K$$
$$\leq M[\phi(\log|f|_\mathscr{C}) + \phi(\log|f_n|_\mathscr{C})] + K$$

σ-a.e. on Δ. Since $\delta < \varepsilon$,

$$\int_\Delta \phi(\log|f - f_n|_\mathscr{C})\,d\sigma < (M + K)\varepsilon + M\int_\Delta \phi(\log|f_n|_\mathscr{C})\,d\sigma$$

by (4-20). By (14-16) and (4-20),

$$\limsup_{n\to\infty} \int_\Delta \phi(\log|f_n|_\mathscr{C})\,d\sigma$$

$$= \lim_{n\to\infty} \int_\Gamma \phi(\log|f_n|_\mathscr{C})\,d\sigma - \liminf_{n\to\infty} \int_{\Gamma\backslash\Delta} \phi(\log|f_n|_\mathscr{C})\,d\sigma$$

$$\leq \int_\Gamma \phi(\log|f|_\mathscr{C})\,d\sigma - \int_{\Gamma\backslash\Delta} \phi(\log|f|_\mathscr{C})\,d\sigma$$

$$= \int_\Delta \phi(\log|f|_\mathscr{C})\,d\sigma$$

$$< \varepsilon.$$

Therefore for all sufficiently large n,

$$\int_\Delta \phi(\log|f - f_n|_\mathscr{C})\,d\sigma < (2M + K)\varepsilon.$$

By the arbitrariness of ε,

$$\lim_{n \to \infty} \int_\Delta \phi(\log|f - f_n|_\mathscr{C}) \, d\sigma = 0. \tag{4-22}$$

Combining (4-21) and (4-22), we obtain (4-17) first for r tending to 1 through the sequence $r_n \uparrow 1$ and then as asserted by the arbitrariness of the sequence $r_n \uparrow 1$. This completes the proof. ∎

DEFINITION.   Let $X = \mathscr{C}$ or $\mathscr{B}(\mathscr{C})$. We write $N_X(\Gamma)$, $H_X^p(\Gamma), \ldots$ for the classes of boundary functions of functions in $N_X(D)$, $H_X^p(D), \ldots$. We define $N_X(R)$, $H_X^p(R), \ldots$ similarly from $N_X(\Pi)$, $H_X^p(\Pi), \ldots$.

The mapping $f(z) \to f(e^{i\theta})$ from $N_X(D)$ to $N_X(\Gamma)$ is one-to-one and linear. If $F(z)$, $G(z) \in N_{\mathscr{B}(\mathscr{C})}(D)$ and $f(z) \in N_\mathscr{C}(D)$, then $F(z)G(z)$ is in $N_{\mathscr{B}(\mathscr{C})}(D)$ and has boundary function $F(e^{i\theta})G(e^{i\theta})$, and $F(z)f(z)$ is in $N_\mathscr{C}(D)$ and has boundary function $F(e^{i\theta})f(e^{i\theta})$. We caution that there is more here than meets the eye since multiplication is not continuous in the strong operator topology. However, by Section 4.3, Theorem A, we can reduce the assertions to the case of bounded functions, and then they are easily proved. The situation for $N_X(\Pi)$ and $N_X(R)$ is similar.

4.7   Hardy Classes on the Disk

It is now easy to derive the main facts concerning boundary behavior for the Hardy classes.

THEOREM A.   Let $X = \mathscr{C}$ or $\mathscr{B}(\mathscr{C})$.

(i) For $0 < p \leq \infty$, $H_X^p(\Gamma) = N_X^+(\Gamma) \cap L_X^p(\sigma)$.
(ii) If $0 < p < \infty$ and $f \in H_X^p(D)$, then

$$\|f\|_p^p = \lim_{r \uparrow 1} \int_\Gamma |f(re^{i\theta})|_X^p \, d\sigma = \int_\Gamma |f(e^{i\theta})|_X^p \, d\sigma.$$

(iii) If $f \in H_X^\infty(D)$, then

$$\|f\|_\infty = \lim_{r \uparrow 1} \max_{|z|=r} |f(z)|_X = \operatorname{ess\,sup}_\Gamma |f(e^{i\theta})|_X.$$

Proof.   (i) For $0 < p < \infty$ this follows from Section 4.6, Theorem C(i), with $\phi(t) = e^{pt}$. The inclusion $H_X^\infty(\Gamma) \subseteq N_X^+(\Gamma) \cap L_X^p(\sigma)$ is clear, and we obtain the reverse inclusion from the result in no. 3 in the Examples and Addenda at the end of the chapter.

(ii) Apply Theorem C(i) of Section 4.6 with $\phi(t) = e^{pt}$.

(iii) The first equality follows from the definition of $\|f\|_\infty$ and the maximum modulus principle (Hille and Phillips [1957], p. 100). Another application of the result in the Examples and Addenda, no. 3, yields

$$\|f\|_\infty \leq \operatorname{ess\,sup}_\Gamma |f(e^{i\theta})|_X.$$

This inequality is also a consequence of the Poisson representation in Theorem C below. The reverse inequality follows from (4-9). ∎

THEOREM B. *If* $f \in H_{\mathscr{C}}^p(D)$, $0 < p < \infty$, *then*

$$\lim_{r \uparrow 1} \int_{\Gamma} |f(e^{i\theta}) - f(re^{i\theta})|_{\mathscr{C}}^p \, d\sigma = 0.$$

Proof. Apply Theorem C(ii) of Section 4.6 with $\phi(t) = e^{pt}$. ∎

THEOREM C. *Let* $X = \mathscr{C}$ *or* $\mathscr{B}(\mathscr{C})$.

(i) *If* $1 \le p \le \infty$, *then* $H_X^p(\Gamma)$ *is the subspace of* $L_X^p(\sigma)$ *consisting of all functions* $f(e^{i\theta})$ *in* $L_X^p(\sigma)$ *such that*

$$\int_{\Gamma} f(e^{i\theta}) e^{ij\theta} \, d\sigma = 0, \qquad j = 1,2,3,\ldots. \tag{4-23}$$

(ii) *If* $f \in H_X^p(D)$, $1 \le p \le \infty$, *then for all* $z \in D$,

$$f(z) = \int_{\Gamma} \frac{f(e^{it})}{1 - ze^{-it}} \, d\sigma = \int_{\Gamma} P(z,e^{it}) f(e^{it}) \, d\sigma. \tag{4-24}$$

The integrals in (4-23) and (4-24) are taken in the weak sense (see Section 4.5). The two formulas for $f(z)$ in (4-24) are called the *Cauchy* and *Poisson representations*, respectively.

Proof. We take the result as known in the scalar case (Duren [1970], Hoffman [1962]). If $f(z)$ belongs to $H_X^p(D)$, $1 \le p \le \infty$, then the boundary function $f(e^{i\theta})$ belongs to $L_X^p(\sigma)$ by Theorem A. Then (4-23) and (4-24) follow by applying the scalar version of the theorem to the functions $\langle f(z),c \rangle_{\mathscr{C}}$ when $X = \mathscr{C}$ and $\langle f(z)a,b \rangle_{\mathscr{C}}$ when $X = \mathscr{B}(\mathscr{C})$ for arbitrary $a,b,c \in \mathscr{C}$.

Conversely, let $f(e^{i\theta})$ be a given function in $L_X^p(\sigma)$ satisfying (4-23). Then since

$$\frac{1}{1 - ze^{-it}} = P(z,e^{it}) - \frac{\bar{z}e^{it}}{1 - \bar{z}e^{it}} = P(z,e^{it}) - \sum_1^\infty \bar{z}^j e^{ijt},$$

there is a function $f(z)$ on D satisfying (4-24). The first representation of $f(z)$ in (4-24) shows that $f(z)$ is holomorphic, and by familiar properties of the Poisson kernel the second implies that $f(z)$ is in $H_X^p(D)$. It remains to show that the boundary function of $f(z)$ is the given function $f(e^{i\theta})$. For definiteness suppose that $X = \mathscr{C}$. It is enough to show that for a countable dense set of vectors $c \in \mathscr{C}$, the boundary function of $\langle f(z),c \rangle_{\mathscr{C}}$ is equal to $\langle f(e^{i\theta}),c \rangle_{\mathscr{C}}$ σ-a.e. This follows from the scalar version of the theorem. The case $X = \mathscr{B}(\mathscr{C})$ is treated similarly. ∎

THEOREM D.   Let $X = \mathscr{C}$ or $\mathscr{B}(\mathscr{C})$. If $1 \leq p \leq \infty$, then $H^p_X(D)$ and $H^p_X(\Gamma)$ are Banach spaces (the norm in $H^p_X(\Gamma)$ is that of $L^p_X(\sigma)$). The mapping $f(z) \to f(e^{i\theta})$ is an isometry from $H^p_X(D)$ onto $H^p_X(\Gamma)$.

Proof.   This result may be obtained as a corollary of Theorem C.   ■

4.8   Hardy Classes on the Half-Plane

The scalar theory for the half-plane is given in Duren [1970], Dym and McKean [1972], Hoffman [1962], and Krylov [1939]. As in the disk case, vector and operator generalizations of many theorems follow in a straightforward way from the classical theory and results of Sections 4.4–4.6. The results stated below present no unusual difficulties, and the proofs can be safely omitted.

THEOREM A.   Let $X = \mathscr{C}$ or $\mathscr{B}(\mathscr{C})$. If $g \in \mathscr{H}^p_X(\Pi)$, $1 \leq p \leq \infty$, then

$$g(z) = \frac{y}{\pi} \int_{-\infty}^{\infty} \frac{g(t)}{(t-x)^2 + y^2} dt, \qquad y > 0. \tag{4-25}$$

THEOREM B.   Let $X = \mathscr{C}$ or $\mathscr{B}(\mathscr{C})$ and fix $p \in [1,\infty)$.

(i) If $g(z)$ is in $H^p_X(\Pi)$, then the boundary function $g(x)$ belongs to $L^p_X(-\infty,\infty)$,

$$g(z) = \frac{1}{2\pi i} \int_{-\infty}^{\infty} \frac{g(t)}{t-z} dt, \qquad y > 0, \tag{4-26}$$

and

$$0 = \frac{1}{2\pi i} \int_{-\infty}^{\infty} \frac{g(t)}{t-z} dt, \qquad y < 0. \tag{4-27}$$

(ii) Conversely let $g(x)$ be a given function in $L^p_X(-\infty, \infty)$ which satisfies (4-27). Then (4-25) and (4-26) define one and the same function $g(z)$ on Π, $g(z)$ belongs to $H^p_X(\Pi)$, and its boundary function is the given function $g(x)$.

THEOREM C.   Let $X = \mathscr{C}$ or $\mathscr{B}(\mathscr{C})$ and fix $p \in [1,\infty]$. Then $H^p_X(\Pi)$ and $H^p_X(R)$ are Banach spaces, and the mapping $g(z) \to g(x)$ is an isometry from $H^p_X(\Pi)$ onto $H^p_X(R)$.

The norm in $H^p_X(R)$ is that of $L^p_X(-\infty,\infty)$. A similar result holds for $\mathscr{H}^p_X(\Pi)$ and $\mathscr{H}^p_X(R)$, provided that the norm in $\mathscr{H}^p_X(R)$ is defined by

$$\||g\||_p = \left(\frac{1}{\pi} \int_{-\infty}^{\infty} \frac{|g(x)|^p_X}{1+x^2} dx \right)^{1/p}$$

if $1 \le p < \infty$ and

$$\||g\||_\infty = \operatorname*{ess\,sup}_{-\infty < x < \infty} |g(x)|_X$$

if $p = \infty$.

The case $p = 2$ and $X = \mathscr{C}$ is especially tractable, thanks to the Plancherel theorem and Paley-Wiener representation.

THEOREM D (*Plancherel Theorem*).   *There is an isometry* $\mathscr{F}: F \to \hat{F}$ *of* $L^2_\mathscr{C}(-\infty,\infty)$ *onto itself such that for each* $F \in L^2_\mathscr{C}(-\infty,\infty)$,

$$\hat{F}(x) = \lim_{A \to \infty} \frac{1}{\sqrt{2\pi}} \int_{-A}^{A} e^{-ixt} F(t)\, dt$$

and

$$F(x) = \lim_{A \to \infty} \frac{1}{\sqrt{2\pi}} \int_{-A}^{A} e^{ixt} \hat{F}(t)\, dt$$

with convergence in the metric of $L^2_\mathscr{C}(-\infty,\infty)$.

THEOREM E (*Paley-Wiener Representation*).   *Given* $f \in L^2_\mathscr{C}(0,\infty)$, *define*

$$F(z) = \frac{1}{\sqrt{2\pi}} \int_{0}^{\infty} e^{itz} f(t)\, dt, \qquad y > 0.$$

Then the mapping $U: f \to F$ *is an isometry from* $L^2_\mathscr{C}(0,\infty)$ *onto* $H^2_\mathscr{C}(\Pi)$. *If* f *and* F *are related in this way, then for each* $y \ge 0$,

$$\lim_{A \to \infty} \frac{1}{\sqrt{2\pi}} \int_{-A}^{A} e^{-ixt} F(t + iy)\, dt = \begin{cases} e^{-yx} f(x), & x > 0, \\ 0, & x < 0, \end{cases}$$

with convergence in the metric of $L^2_\mathscr{C}(-\infty, \infty)$.

In particular, $H^2_\mathscr{C}(\Pi)$ is a Hilbert space which is naturally isomorphic with $L^2_\mathscr{C}(0,\infty)$.

THEOREM F.   *For any* $F \in H^2_\mathscr{C}(\Pi)$, $\int_{-\infty}^{\infty} |F(x + iy)|^2_\mathscr{C}\, dx$ *is a nonincreasing function of* $y > 0$. *For any* $F, G \in H^2_\mathscr{C}(\Pi)$,

$$\langle F, G \rangle_2 = \lim_{y \downarrow 0} \int_{-\infty}^{\infty} \langle F(x + iy), G(x + iy) \rangle_\mathscr{C}\, dx$$

and

$$\lim_{y \downarrow 0} \int_{-\infty}^{\infty} |F(x + iy) - F(x)|^2_\mathscr{C}\, dx = 0.$$

Examples and Addenda

1. A function $F \in H^{\infty}_{\mathscr{B}(\mathscr{C})}(D)$ may in every direction fail to have a radial limit in the operator norm. An example is

$$F(z) = \sum_{0}^{\infty} \langle \cdot, e_j \rangle_{\mathscr{C}} e_j z^j, \qquad z \in D,$$

where $\{e_j\}_0^{\infty}$ is an orthonormal sequence in $\mathscr{C}$ (dim $\mathscr{C} = \infty$). This function also shows that the assertion of Theorem C(ii) in Section 4.6 fails if $\mathscr{C}$ is replaced by $\mathscr{B}(\mathscr{C})$.

2. If dim $\mathscr{C} < \infty$, then every holomorphic $\mathscr{B}(\mathscr{C})$-valued function F such that Im $F \geq 0$ on Π belongs to $N^+_{\mathscr{B}(\mathscr{C})}(\Pi)$. However, this is not true if dim $\mathscr{C} = \infty$. For an example, let F be the resolvent of the self-adjoint operator multiplication by x on $L^2(-\infty,\infty)$. Then $\lim_{y \downarrow 0} F(x + iy)$ does not exist in the weak operator topology for any real x. Such a function cannot belong even to $N_{\mathscr{B}(\mathscr{C})}(\Pi)$.

3. Let $f \equiv 0$ be a holomorphic X-valued function on D, where $X = \mathscr{C}$ or $\mathscr{B}(\mathscr{C})$. Then $f \in N^+_X(D)$ if and only if

$$\log|f(z)|_X \leq \int_{\Gamma} P(z, e^{it}) \log w(e^{it}) \, d\sigma, \qquad z \in D,$$

for some positive function w such that $\log w \in L^1(\sigma)$. In this case, the inequality holds with $w(e^{it}) = |f(e^{it})|_X$ σ-a.e.

4. Let $f \not\equiv 0$ be a holomorphic X-valued function on Π, where $X = \mathscr{C}$ or $\mathscr{B}(\mathscr{C})$. Then $f \in N^+_X(\Pi)$ if and only if

$$\log|f(z)|_X \leq \frac{y}{\pi} \int_{-\infty}^{\infty} \frac{\log w(t)}{(t-x)^2 + y^2} \, dt, \qquad z \in \Pi,$$

for some positive function $w(t)$ such that $(1 + t^2)^{-1} \log w(t) \in L^1(-\infty,\infty)$. In this case, the inequality holds with $w(t) = |f(t)|_X$ a.e.

5. For each $c \in \mathscr{C}$ and $w \in \Pi$, $c/[2\pi i(\bar{w} - z)]$ belongs to $H^2_{\mathscr{C}}(\Pi)$ as a function of z, and for each $F(z)$ in $H^2_{\mathscr{C}}(\Pi)$,

$$\langle F(z), c/[2\pi i(\bar{w} - z)] \rangle_2 = \langle F(w), c \rangle_{\mathscr{C}}.$$

A similar result for $H^2_{\mathscr{C}}(D)$ is given in the lemma to Theorem B in Section 1.15.

6. The mapping

$$f(z) \to F(z) = \pi^{-\frac{1}{2}}(z + i)^{-1} f\left(\frac{z-i}{z+i}\right)$$

is a Hilbert space isomorphism from $H^2_{\mathscr{C}}(D)$ onto $H^2_{\mathscr{C}}(\Pi)$.

Notes

The account in Chapter 4 is an expanded version of the approach to Nevanlinna and Hardy classes of vector and operator valued functions in Rosenblum and Rovnyak [1971]. Heins [1969], [1983] uses similar methods and obtains additional results on the Poisson representation and Fatou's theorem for vector valued harmonic functions on the unit disk.

The use of subharmonic functions in the study of Hardy classes is standard in both the classical theory and its generalizations. See, for example, Parreau [1952], Rudin [1955], Heins [1969], Duren [1970], and Hasumi [1983].

Hardy-Orlicz classes on the unit disk D were introduced by Weiss [1956], according to whom the idea was suggested by A. Zygmund. At the same time, we note that similar notions were previously considered by Nevanlinna [1951] and Parreau [1952]. Hardy-Orlicz classes have been studied by Kraynek [1972] and Leśniewicz [1971], [1973I-III].

The theory of vector and operator valued functions is also discussed in Helson [1964], Hille and Phillips [1957], Sz.-Nagy and Foiaş [1970], and Ziegler [1982]. There are many sources for the classical scalar theory, such as Duren [1970], Dym and McKean [1972], Garnett [1981], Hoffman [1962], Koosis [1980], Krylov [1939], and Zygmund [1968].

Chapter 5

Operator Valued Inner and Outer Functions

5.1 Introduction

In this chapter we construct an inner-outer factorization theory for operator valued holomorphic functions that are of bounded type on a disk or half-plane. The theory is less complete than in the scalar case, but it retains many of the characteristic features of the classical situation. In the case of bounded functions we obtain our results from the factorization theory for Toeplitz operators in Chapter 3. Unbounded functions are handled with the aid of scalar mollifiers. In Sections 5.9 and 5.10 we characterize outer functions in terms of extremal properties.

Notation is as in Chapter 4. In particular, $\mathscr{C}$ always denotes a separable Hilbert space, and D and Π are the open unit disk and open upper half-plane.

5.2 Inner and Outer Functions

Let $\Omega = D$ or Π.

DEFINITION A. *If* $A \in H^\infty_{\mathscr{B}(\mathscr{C})}(\Omega)$, *then:*

(i) *A is an* inner function *if the operator*

$$T(A): f \to Af, \qquad f \in H^2_{\mathscr{C}}(\Omega), \tag{5-1}$$

is a partial isometry on $H^2_{\mathscr{C}}(\Omega)$;

(ii) *A is an* outer function *if*

$$\bigvee \{Af: f \in H^2_{\mathscr{C}}(\Omega)\} = H^2_M(\Omega) \tag{5-2}$$

for some subspace M of $\mathscr{C}$.

We use a scalar mollifier to extend the definition of an outer function to allow for unbounded functions.

DEFINITION B. *A holomorphic* $\mathscr{B}(\mathscr{C})$-*valued function F on* Ω *is an* outer function *if there is a bounded scalar valued outer function* $\phi \not\equiv 0$ *on* Ω *such that* ϕF *is bounded and outer in the sense of Definition A.*

The boundary function of an inner (resp. outer) function is also called an inner (resp. outer) function. The function identically zero is both inner and outer by Definition A.

THEOREM A.   *In the scalar case, except for the function identically zero, the classes of inner and outer functions obtained from Definitions A and B coincide with the classes obtained from the classical definitions.*

The classical definitions and properties of inner and outer functions are given in Duren [1970] and Hoffman [1962].

Proof.   For definiteness take $\Omega = D$. In the scalar case, a function A in $H^\infty(D)$, $A \not\equiv 0$, is inner in the sense of Definition A if and only if $\langle Af, Ag \rangle_2 = \langle f, g \rangle_2$, or what is the same thing,

$$\int_\Gamma |A(e^{i\theta})|^2 f(e^{i\theta}) \bar{g}(e^{i\theta}) \, d\sigma = \int_\Gamma f(e^{i\theta}) \bar{g}(e^{i\theta}) \, d\sigma$$

for all f, $g \in H^2(D)$. This holds if and only if $|A(e^{i\theta})| = 1\sigma$-a.e., that is, A is inner in the classical sense.

The equivalence of the two definitions of an outer function follows from Beurling's theorem (Duren [1970], p. 114) in the case of bounded functions. The general case is easily reduced to this case.   ∎

The *canonical shift operators* on $H^2_{\mathscr{C}}(D)$ and $H^2_{\mathscr{C}}(\Pi)$ are defined by

$$S: f(z) \to zf(z) \qquad \text{on } H^2_{\mathscr{C}}(D) \tag{5-3}$$

and

$$S: f(z) \to \frac{z - i}{z + i} f(z) \qquad \text{on } H^2_{\mathscr{C}}(\Pi). \tag{5-4}$$

These operators are unitarily equivalent by means of the isomorphism in the Examples and Addenda to Chapter 4, no. 6. Therefore the results for the operator (5-3) in 1.15 transfer to analogous results for (5-4). In either case, $\Omega = D$ or Π, we set

$$\mathscr{K} = \ker S^* \qquad \text{and} \qquad P_0 = I - SS^*.$$

By the Wold decomposition 1.3, P_0 is the projection of $H^2_{\mathscr{C}}(\Omega)$ on $\mathscr{K}$, and each f in $H^2_{\mathscr{C}}(\Omega)$ has an expansion

$$f = \sum_0^\infty S^j P_0 S^{*j} f \tag{5-5}$$

which converges in the metric of $H^2_{\mathscr{C}}(\Omega)$. The expansion (5-5) also converges pointwise on Ω in the norm of $\mathscr{C}$ and is easily identified in classical terms.

THEOREM B.   (i) *When $\Omega = D$, $\mathscr{K}$ is the space of constant functions in $H^2_{\mathscr{C}}(D)$, that is, with an obvious identification, $\mathscr{K} = \mathscr{C}$. For each $f \in H^2_{\mathscr{C}}(D)$, (5-5) coincides with the Taylor expansion*

$$f(z) = \sum_0^\infty a_j z^j, \qquad z \in D.$$

(ii) *When $\Omega = \Pi$, $\mathcal{K}$ is the space of functions of the form $\pi^{-1/2}c/(z + i)$, where $c \in \mathcal{C}$. For each $f \in H^2_{\mathcal{C}}(\Pi)$, (5-5) coincides with the expansion*

$$f(z) = \sum_0^\infty \frac{\pi^{-1/2}a_j}{z + i}\left(\frac{z - i}{z + i}\right)^j, \qquad z \in \Pi,$$

where $a_0, a_1, \ldots$ are the Taylor coefficients of the function $g \in H^2_{\mathcal{C}}(D)$ such that $f(z) = \pi^{-1/2}(z + i)^{-1}g((z - i)/(z + i))$ for all $z \in \Pi$.

COROLLARY. *A subspace $\mathcal{M}$ of $H^2_{\mathcal{C}}(\Omega)$ reduces the canonical shift operator S if and only if $\mathcal{M} = H^2_M(\Omega)$ for some subspace M of $\mathcal{C}$.*

The next result enables us to translate many theorems on operators, such as the factorization theorems for Toeplitz operators in Chapter 3, into analogous theorems on operator valued functions.

THEOREM C. *Let S be the canonical shift operator on $H^2_{\mathcal{C}}(\Omega)$. Let $A \in H^\infty_{\mathcal{B}(\mathcal{C})}(\Omega)$ and define $T(A)$ on $H^2_{\mathcal{C}}(\Omega)$ by (5-1). Then:*

  (i) *$T(A)$ is S-analytic in the sense of 1.6, and every S-analytic operator on $H^2_{\mathcal{C}}(\Omega)$ has this form;*
 (ii) *$T(A)$ is S-constant in the sense of 1.6 if and only if $A(z) \equiv \text{const. on } \Omega$;*
(iii) *$T(A)$ is S-inner (resp. S-outer) in the sense of 1.6 if and only if A is an inner (resp. outer) function;*
(iv) *$T(A)$ is both S-inner and S-constant if and only if $A(z) \equiv A_0$ on Ω where $A_0 \in \mathcal{B}(\mathcal{C})$ is a partial isometry.*

These results are straightforward, and we omit the proofs.

5.3 The Subspace $M_{in}(A)$

It turns out that any $\mathcal{B}(\mathcal{C})$-valued inner function A does all of its work on a subspace M of $\mathcal{C}$ and is trivial on the orthogonal complement of M. This subspace is denoted $M_{in}(A)$. The formal definition and key properties are given below.

THEOREM A. *Let A be a $\mathcal{B}(\mathcal{C})$-valued inner function on $\Omega = D$ or Π. There exists a subspace M of $\mathcal{C}$ such that*

$$\{f : f \in H^2(\Omega) \quad and \quad \|Af\|_2 = \|f\|_2\} = H^2_M(\Omega)$$

and $Ag = 0$ for all $g \in H^2_N(\Omega)$, where $N = M^\perp$.

Proof. The set of functions f in $H^2_{\mathcal{C}}(\Omega)$ such that $\|Af\|_2 = \|f\|_2$ is a reducing subspace for the canonical shift operator S by Section 1.7, Theorem A. By Section 5.2 (corollary to Theorem B), this subspace has the form $H^2_M(\Omega)$ for some subspace M of $\mathcal{C}$. Moreover, $Ag = 0$ for every $g \in H^2_M(\Omega)^\perp = H^2_N(\Omega)$, where $N = M^\perp$. ∎

DEFINITION. *We write $M_{in}(A)$ for the subspace M in the situation of Theorem A.*

THEOREM B.   *For any $\mathscr{B}(\mathscr{C})$-valued inner function A on $\Omega = D$ or Π,*

$$M_{\text{in}}(A) = \bigvee_{w \in \Omega_0} A(w)^* \mathscr{C}, \tag{5-6}$$

where Ω_0 is any subset of Ω that has an accumulation point in Ω.

Proof.   For definiteness let $\Omega = D$. If $M = M_{\text{in}}(A)$, then $H_M^2(D)$ is the initial space of the partial isometry $T(A)$ defined by (5-1). Equivalently, $H_M^2(D)$ is the range of $T(A)^*$. Functions of the form $c/(1 - z\bar{w})$, $c \in \mathscr{C}$, $w \in \Omega_0$, span a dense subset of $H_{\mathscr{C}}^2(D)$, and

$$T(A)^*: c/(1 - z\bar{w}) \to A(w)^* c/(1 - z\bar{w}).$$

Thus (5-6) follows.   ∎

THEOREM C.   *Let A be a $\mathscr{B}(\mathscr{C})$-valued inner function $\Omega = D$ or Π. Then the values of the nontangential boundary function of A are partial isometries on $\mathscr{C}$ with initial space $M = M_{\text{in}}(A)$ σ-a.e. on Γ or a.e. on $(-\infty,\infty)$, depending on the case.*

A converse result is given in the Examples and Addenda, no. 1.

Proof.   It is sufficient to prove this when $\Omega = D$. Let P_M be the projection of $\mathscr{C}$ on M. Since the operator (5-1) is a partial isometry with initial space $H_M^2(D)$,

$$\int_\Gamma \langle A(e^{i\theta})^* A(e^{i\theta}) a, b \rangle_{\mathscr{C}} e^{i(j-k)\theta} \, d\sigma$$

$$= \langle z^j A(z) a, z^k A(z) b \rangle_2 = \langle z^j P_M a, z^k P_M b \rangle_2$$

$$= \int_\Gamma \langle P_M a, b \rangle_{\mathscr{C}} e^{i(j-k)\theta} \, d\sigma$$

for all $a, b \in \mathscr{C}$, $j, k = 0, 1, 2, \ldots$. Hence $A(e^{i\theta})^* A(e^{i\theta}) = P_M$ σ-a.e. on Γ, and the result follows.   ∎

5.4   The Subspace $M_{\text{out}}(F)$

We show that the values of any $\mathscr{B}(\mathscr{C})$-valued outer function F on D or Π have ranges that are dense in a constant subspace M of $\mathscr{C}$. This subspace is denoted $M_{\text{out}}(F)$. The formal definition follows a preliminary result. Except where otherwise stated, we assume that $\Omega = D$ or Π.

THEOREM A.   *Let F be a $\mathscr{B}(\mathscr{C})$-valued holomorphic function on Ω, and let $\phi \not\equiv 0$ and $\psi \not\equiv 0$ be bounded scalar valued holomorphic functions such that ϕF and ψF are bounded on Ω. If ϕ and ψ are outer, then*

$$(\phi F H_{\mathscr{C}}^2(\Omega))^- = (\psi F H_{\mathscr{C}}^2(\Omega))^-. \tag{5-7}$$

Proof. Take $\Omega = D$. By Beurling's theorem (Duren [1970], p. 114), $(\phi H^2(D))^- = (\psi H^2(D))^- = H^2(D)$. A routine approximation argument yields $(\phi H_{\mathscr{C}}^2(D))^- = (\psi H_{\mathscr{C}}^2(D))^- = H_{\mathscr{C}}^2(D)$. Hence

$$(\phi F H_{\mathscr{C}}^2(D))^- = (\phi F(\psi H_{\mathscr{C}}^2(D))^-)^- = (\phi \psi F H_{\mathscr{C}}^2(D))^-,$$

and similarly with the roles of ϕ and ψ interchanged. Thus (5-7) follows. ∎

DEFINITION.   *Let F be a $\mathscr{B}(\mathscr{C})$-valued outer function on Ω, and let $\phi \not\equiv 0$ be a bounded scalar valued outer function such that ϕF is bounded on Ω. Then*

$$(\phi F H_{\mathscr{C}}^2(\Omega))^- = H_M^2(\Omega), \tag{5-8}$$

where M is a subspace of $\mathscr{C}$ that does not depend on the choice of ϕ by Theorem A. We set $M_{\text{out}}(F) = M$ in this situation.

THEOREM B.   *For any $\mathscr{B}(\mathscr{C})$-valued outer function F on Ω and any $w \in \Omega$,*

$$M_{\text{out}}(F) = (F(w)\mathscr{C})^-. \tag{5-9}$$

Proof. We may assume without loss of generality that $\Omega = D$ and F is bounded. Let $M = M_{\text{out}}(F)$, so

$$(F H_{\mathscr{C}}^2(D))^- = H_M^2(D). \tag{5-10}$$

For any $c \in \mathscr{C}$, we have $c \perp M$ in the inner product of $\mathscr{C}$ if and only if $c \perp H_M^2(D)$ in the inner product of $H_{\mathscr{C}}^2(D)$ (in the second relation, c is viewed as a constant function on D). By (5-10), this holds if and only if for all $f \in H_{\mathscr{C}}^2(D)$,

$$\langle f(0), F(0)^*c \rangle_{\mathscr{C}} = \langle F(z)f(z), c \rangle_2 = 0,$$

that is, $F(0)^*c = 0$. Hence (5-9) holds for $w = 0$.

For any $w \in D$, set

$$\delta_w(z) = (w - z)/(1 - z\bar{w}), \qquad z \in D.$$

Then $C_w : f \to f \circ \delta_w$ is an invertible operator on $H_{\mathscr{C}}^2(D)$. Set $F_w = F \circ \delta_w$. Then

$$(F_w H_{\mathscr{C}}^2(D))^- = \{C_w(F H_{\mathscr{C}}^2(D))\}^- = C_w\{(F H_{\mathscr{C}}^2(D))^-\}$$
$$= C_w H_M^2(D) = H_M^2(D).$$

Therefore F_w is outer and $M_{\text{out}}(F_w) = M = M_{\text{out}}(F)$. By what we just proved,

$$(F(w)\mathscr{C})^- = (F_w(0)\mathscr{C})^- = M_{\text{out}}(F_w) = M_{\text{out}}(F),$$

and (5-9) holds. ∎

5.5   The Beurling-Lax Theorem, Revisited

The Beurling-Lax theorem 1.12 is usually stated in terms of inner functions.

THEOREM A *(Beurling [1949], Lax [1959]).*   *A subspace $\mathscr{M}$ of $H_{\mathscr{C}}^2(\Omega)$, $\Omega = D$ or Π, is invariant under the canonical shift operator S if and only if $\mathscr{M} = A H_{\mathscr{C}}^2(\Omega)$ for some $\mathscr{B}(\mathscr{C})$-valued inner function A on Ω.*

By Section 5.2, Theorem C, this is a special case of 1.12. Similarly, the assertions following 1.12 yield a companion uniqueness result.

THEOREM B. *If $AH^2_{\mathscr{C}}(\Omega) = CH^2_{\mathscr{C}}(\Omega)$ for two inner functions A, C on Ω $(\Omega = D$ or $\Pi)$, then*

$$C(z) \equiv A(z)B_0, \qquad A(z) \equiv C(z)B_0^* \text{ on } \Omega,$$

where $B_0 \in \mathscr{B}(\mathscr{C})$ is a partial isometry with initial space $M_{\text{in}}(C)$ and final space $M_{\text{in}}(A)$. Conversely, if two inner functions A, C on Ω are so related, then $AH^2_{\mathscr{C}}(\Omega) = CH^2_{\mathscr{C}}(\Omega)$.

5.6 Theorem. Canonical Factorization of Functions of Bounded Type

Let $\Omega = D$ or Π. Every $F \in N_{\mathscr{B}(\mathscr{C})}(\Omega)$ has a representation

$$F = AG/b, \tag{5-11}$$

where A is a $\mathscr{B}(\mathscr{C})$-valued inner function, G is a $\mathscr{B}(\mathscr{C})$-valued outer function with $M_{\text{out}}(G) = M_{\text{in}}(A)$, and b is a scalar valued singular inner function. For any representation (5-11), either

$$F(e^{i\theta})^*F(e^{i\theta}) = G(e^{i\theta})^*G(e^{i\theta}) \quad \sigma\text{-a.e. on } \Gamma \tag{5-12}$$

or

$$F(x)^*F(x) = G(x)^*G(x) \quad \text{a.e. on } (-\infty,\infty), \tag{5-13}$$

depending on the case. For any $w_0 \in \Omega$, there exists a representation (5-11) such that $G(w_0) \geq 0$.

Proof. We take $\Omega = D$ and $w_0 = 0$. The general case follows by conformal mapping. For any $C \in H^\infty_{\mathscr{B}(\mathscr{C})}(D)$, let $T(C)$ be the operator multiplication by C on $H^2_{\mathscr{C}}(D)$.

Suppose first that $F \in H^\infty_{\mathscr{B}(\mathscr{C})}(D)$. By 3.6 and 5.2, Theorem C,

$$T(F) = T(A)T(G)$$

for some inner function A and some bounded outer function G such that $M_{\text{in}}(A) = M_{\text{out}}(G)$. Then $F = AG$, so there exists a factorization (5-11) with $b \equiv 1$. Moreover, by 3.6 we can choose the factorization so that for all $c \in \mathscr{C}$,

$$\langle G(0)c,c\rangle_{\mathscr{C}} = \langle G(z)c,c\rangle_2 \geq 0,$$

that is, $G(0) \geq 0$.

Now consider any $F \in N_{\mathscr{B}(\mathscr{C})}(D)$. By 4.3, Theorem A, $F = F_0/u$, where $F_0 \in H^\infty_{\mathscr{B}(\mathscr{C})}(D)$ and u is a scalar valued holomorphic function such that $0 < |u| \leq 1$ on D. Factor $F_0 = A_0 G_0$ as above with $G_0(0) \geq 0$. Factor $u = bv$, where b is inner and v is outer with $v(0) > 0$. Since $u \neq 0$ on D, b is a singular inner function. The required factorization (5-11) is then obtained with $A = A_0$, $G = G_0/v$, and the singular inner function b.

We prove (5-12) for any factorization (5-11). First let $F,G \in H^\infty_{\mathscr{B}(\mathscr{C})}(D)$ and $b \equiv 1$. By 3.6,

$$T(F)^*T(F) = T(G)^*T(G). \tag{5-14}$$

Hence for any $f_1,f_2 \in H^2_{\mathscr{C}}(D)$,

$$\langle Ff_1,Ff_2 \rangle_2 = \langle Gf_1,Gf_2 \rangle_2, \tag{5-15}$$

and so

$$\int_\Gamma \langle F(e^{i\theta})^*F(e^{i\theta})f_1(e^{i\theta}),f_2(e^{i\theta}) \rangle_{\mathscr{C}} \, d\sigma$$

$$= \int_\Gamma \langle G(e^{i\theta})^*G(e^{i\theta})f_1(e^{i\theta}),f_2(e^{i\theta}) \rangle_{\mathscr{C}} \, d\sigma. \tag{5-16}$$

By the arbitrariness of f_1,f_2, (5-12) follows.

The general case of (5-12) can be reduced to the special case. Consider any factorization (5-11). By 4.3, Theorem A, $F = F_0/(u_iu_o)$, where $F_0 \in H^\infty_{\mathscr{B}(\mathscr{C})}(D)$, u_i, u_o are scalar valued functions, u_i is a singular inner function, and u_o is an outer function such that $0 < |u_o| \le 1$ on D. By 5.2, $G = G_0/v$, where G_0 is a bounded outer function and v is a bounded scalar valued outer function on D. By (5-11),

$$bvF_0 = (u_iA)(u_0G_0).$$

Applying the special case to this factorization, we get

$$[b(e^{i\theta})v(e^{i\theta})F_0(e^{i\theta})]^*[b(e^{i\theta})v(e^{i\theta})F_0(e^{i\theta})]$$

$$= [u_o(e^{i\theta})G_0(e^{i\theta})]^*[u_o(e^{i\theta})G_0(e^{i\theta})]$$

σ-a.e. on Γ, which implies (5-12).   ∎

5.7   Theorem. Canonical Factorization of Functions of Class N^+

Let $\Omega = D$ or Π. Every $F \in N^+_{\mathscr{B}(\mathscr{C})}(\Omega)$ has a representation

$$F = AG, \tag{5-17}$$

where A is a $\mathscr{B}(\mathscr{C})$-valued inner function and G is a $\mathscr{B}(\mathscr{C})$-valued outer function such that $M_{\mathrm{out}}(G) = M_{\mathrm{in}}(A)$. For any representation (5-17) either (5-12) or (5-13) holds, depending on the case. For any $w_0 \in \Omega$, there exists a representation (5-17) such that $G(w_0) \ge 0$.

Proof.   The argument is essentially the same as for 5.6. In place of Section 4.3, Theorem A, use Section 4.3, Theorem C.   ∎

COROLLARY.   *If* $\Omega = D$ *or* Π, $N^+_{\mathscr{B}(\mathscr{C})}(\Omega)$ *is the smallest algebra containing all* $\mathscr{B}(\mathscr{C})$-*valued inner and outer functions on* Ω.

5.8   Uniqueness of Inner-Outer Factorizations

Let $F,G \in N^+_{\mathscr{B}(\mathscr{C})}(\Omega)$, $\Omega = D$ or Π.

THEOREM A.   *If G is outer, the following are equivalent:*

(i) *the boundary functions of F and G satisfy (5-12) or (5-13), depending on the case;*

(ii) *$F = AG$ for some $\mathscr{B}(\mathscr{C})$-valued inner function A such that $M_{in}(A) = M_{out}(G)$.*

Proof.   (i) $\Rightarrow$ (ii) For definiteness take $\Omega = D$. Using 4.3, Theorem C, we easily reduce to the case in which F and G are bounded. In this case multiplication by F and multiplication by G are bounded operators $T(F)$ and $T(G)$ on $H^2_{\mathscr{C}}(D)$. Our hypothesis (5-12) implies (5-16) for arbitrary $f_1, f_2 \in H^2_{\mathscr{C}}(D)$. Hence (5-15) and (5-14) hold, and (ii) follows from 3.5, Theorem A.

(ii) $\Rightarrow$ (i) This follows from 5.7.   ∎

THEOREM B.   *If F,G are both outer, the following are equivalent:*

(i) *the boundary functions of F and G satisfy (5-12) or (5-13), depending on the case;*

(ii) *$G(z) \equiv CF(z)$ and $F(z) \equiv C^*G(z)$ on Ω, where $C \in \mathscr{B}(\mathscr{C})$ is a partial isometry with initial space $M_{out}(F)$ and final space $M_{out}(G)$.*

Proof.   Argue as in the proof of Theorem A, but in place of 3.5, Theorem A use the corollary to 3.5, Theorem A.   ∎

THEOREM C.   *Let F,G both be outer, and let their boundary functions satisfy (5-12) or (5-13), depending on the case. If $F(w_0) \geq 0$ and $G(w_0) \geq 0$ for some $w_0 \in \Omega$, then $F \equiv G$ on Ω.*

Proof.   Without loss of generality we can take $\Omega = D$ and $w_0 = 0$. It is easy to reduce the assertion to the case where F and G are bounded, and then the result follows from 3.5, Theorem B.   ∎

5.9   Outer Functions on the Disk

Throughout this section S denotes the canonical shift operator on $H^2_{\mathscr{C}}(D)$. We identify $\mathscr{K} = \ker S^*$ with $\mathscr{C}$ in the obvious way. The Taylor coefficients of $\mathscr{B}(\mathscr{C})$-valued holomorphic functions $A, B, \ldots$ on the disk are denoted $\{A_j\}^\infty_0, \{B_j\}^\infty_0, \ldots$. If $A \in H^\infty_{\mathscr{B}(\mathscr{C})}(D)$, then $T(A)$ is the operator multiplication by A on $H^2_{\mathscr{C}}(D)$. The matrix of $T(A)$ as defined in 3.2 is given by

$$
T(A) \sim \begin{bmatrix} A_0 & 0 & 0 & \cdots \\ A_1 & A_0 & 0 & \cdots \\ A_2 & A_1 & A_0 & \cdots \\ & \cdots & & \end{bmatrix}.
$$

THEOREM A.   *Let* $C \in N_{\mathscr{B}(\mathscr{C})}^{+}(D)$. *Then* C *is outer if and only if*

$$C_0^* C_0 \geq A_0^* A_0 \tag{5-18}$$

for every $A \in N_{\mathscr{B}(\mathscr{C})}^{+}(D)$ *such that*

$$A(e^{i\theta})^* A(e^{i\theta}) = C(e^{i\theta})^* C(e^{i\theta}) \quad \sigma\text{-a.e. on } \Gamma. \tag{5-19}$$

In this case, for every $A \in N_{\mathscr{B}(\mathscr{C})}^{+}(D)$ *satisfying* (5-19), *we have*

$$\sum_0^n C_j^* C_j \geq \sum_0^n A_j^* A_j, \qquad n = 0,1,2,\dots. \tag{5-20}$$

Proof.   If we replace $N_{\mathscr{B}(\mathscr{C})}^{+}(D)$ by $H_{\mathscr{B}(\mathscr{C})}^{\infty}(D)$, the theorem follows from 3.10 (see Theorem C and the corollary to Theorem A). We deduce the general result from the bounded version.

Consider any $C \in N_{\mathscr{B}(\mathscr{C})}^{+}(D)$. By 4.3, Theorem C, there is a scalar valued outer function v such that $v(0) > 0, 0 < |v| \leq 1$, and $C' = vC$ is bounded on D. For each $k = 1,2,3,\dots$, the function defined on D by

$$v_k(z) = \exp\left(\frac{1}{2\pi} \int_0^{2\pi} \frac{e^{i\theta} + z}{e^{i\theta} - z} \log(\min(k|v(e^{i\theta})|,1)) \, d\sigma\right)$$

is outer, $0 < |v_k| \leq 1$, and $C^{(k)} = v_k C$ is bounded on D. Moreover, $v_k(z) \to 1$ uniformly on all compact subsets of D. Thus if

$$\frac{1}{v_k(z)} = \sum_{j=0}^{\infty} \alpha_{jk} z^j, \qquad z \in D,$$

then

$$\lim_{k \to \infty} \alpha_{jk} = \begin{cases} 1, & j = 0, \\ 0, & j \geq 1. \end{cases} \tag{5-21}$$

Suppose that C is outer. For any $A \in N_{\mathscr{B}(\mathscr{C})}^{+}(D)$ satisfying (5-19), set $A^{(k)} = v_k A$, $k \geq 1$. For $k \geq 1$,

$$A^{(k)}(e^{i\theta})^* A^{(k)}(e^{i\theta}) = C^{(k)}(e^{i\theta})^* C^{(k)}(e^{i\theta})$$

σ-a.e. on Γ. Since $C^{(k)}$ is bounded on D, so is $A^{(k)}$ by 4.7, Theorem A. By the special case of the theorem noted above,

$$\sum_{j=0}^n C_j^{(k)*} C_j^{(k)} \geq \sum_{j=0}^n A_j^{(k)*} A_j^{(k)}, \qquad n = 0,1,2,\dots.$$

By (5-21), (5-20) follows on letting $k \to \infty$. In particular, (5-18) holds.

Conversely, suppose that (5-18) holds for every $A \in N_{\mathscr{B}(\mathscr{C})}^{+}(D)$ that satisfies (5-19). Consider any $A' \in N_{\mathscr{B}(\mathscr{C})}^{+}(D)$ such that

$$A'(e^{i\theta})^* A'(e^{i\theta}) = C'(e^{i\theta})^* C'(e^{i\theta})$$

σ- a.e. on Γ. By 4.7, Theorem A, $A' \in H^\infty_{\mathcal{B}(\mathscr{C})}(D)$. If $A = A'/v$, then $A \in N^+_{\mathcal{B}(\mathscr{C})}(D)$ and (5-19) holds. Then by assumption (5-18) holds, so

$$C_0'^* C_0' = |v(0)|^2 C_0^* C_0 \geq |v(0)|^2 A_0^* A_0 = A_0'^* A_0'.$$

Since the result is known for bounded functions, C' is outer. Hence C is outer. This completes the proof. ∎

Let $\mathscr{P}_\mathscr{C}$ be the set of polynomials $p(z) = p_0 + p_1 z + \cdots + p_n z^n$ with coefficients in $\mathscr{C}$.

THEOREM B. *Let $C \in N^+_{\mathcal{B}(\mathscr{C})}(D)$, and assume that $C(z)a$ belongs to $H^2_\mathscr{C}(D)$ for each $a \in \mathscr{C}$. Then C is outer if and only if for all $a \in \mathscr{C}$,*

$$\langle C_0^* C_0 a, a \rangle_\mathscr{C} = \inf_{p \in \mathscr{P}_\mathscr{C}} \int_\Gamma \langle C(e^{i\theta})^* C(e^{i\theta})[a - e^{i\theta} p(e^{i\theta})], a - e^{i\theta} p(e^{i\theta}) \rangle_\mathscr{C} \, d\sigma. \quad (5\text{-}22)$$

In this case, for all $a \in \mathscr{C}$ and $n = 0,1,2,\ldots,$

$$\left\langle \sum_0^n C_j^* C_j a, a \right\rangle_\mathscr{C}$$

$$= \inf_{p \in \mathscr{P}_\mathscr{C}} \int_\Gamma \langle C(e^{i\theta})^* C(e^{i\theta})[a - e^{i(n+1)\theta} p(e^{i\theta})], a - e^{i(n+1)\theta} p(e^{i\theta}) \rangle_\mathscr{C} \, d\sigma. \quad (5\text{-}23)$$

The infimum in (5-22) may be viewed as a form of Szegö's infimum (Ahiezer [1956], Grenander and Szegö [1958], and Dym and McKean [1972]).

Proof. Assume that C is outer. The proof of (5-23) is similar to the proof of 3.10, Theorem C, (i) $\Rightarrow$ (iii). Fix $a \in \mathscr{C}$ and $n \geq 0$. Let $M = M_{\text{out}}(C)$, so $\{Cp : p \in \mathscr{P}_\mathscr{C}\}^- = H^2_M(D)$. By Section 1.2 (Corollary 2), the infimum of $\|Ca - S^{n+1}g\|_2^2$ over all $g \in H^2_M(D)$ is attained with $g = S^{*n+1}Ca$. Thus

$$\inf_{p \in \mathscr{P}_\mathscr{C}} \int_\Gamma \langle C(e^{i\theta})^* C(e^{i\theta})[a - e^{i(n+1)\theta} p(e^{i\theta})], a - e^{i(n+1)\theta} p(e^{i\theta}) \rangle_\mathscr{C} \, d\sigma$$

$$= \inf_{p \in \mathscr{P}_\mathscr{C}} \|Ca - S^{n+1}Cp\|_2^2$$

$$= \inf_{h \in H^2_M(D)} \|Ca - S^{n+1}h\|_2^2$$

$$= \|Ca - S^{n+1}S^{*n+1}Ca\|_2^2$$

$$= \left\langle \sum_0^n C_j^* C_j a, a \right\rangle_\mathscr{C}$$

This proves (5-23), and (5-22) follows as a special case.

Conversely, assume that C satisfies (5-22) for all $a \in \mathscr{C}$. We apply Theorem A to show that C is outer. Let $A \in N^+_{\mathcal{B}(\mathscr{C})}(D)$, and suppose that (5-19) holds. For any

$a \in \mathscr{C}$, by (5-22) and (5-19),

$$\langle C_0^* C_0 a, a \rangle_{\mathscr{C}} = \inf_{p \in \mathscr{P}_{\mathscr{C}}} \int_\Gamma \langle A(e^{i\theta})^* A(e^{i\theta})[a - e^{i\theta}p(e^{i\theta})], a - e^{i\theta}p(e^{i\theta})\rangle_{\mathscr{C}} \, d\sigma$$

$$= \inf_{p \in \mathscr{P}_{\mathscr{C}}} \|A(z)[a - zp(z)]\|_2^2$$

$$\geq |A(0)a|_{\mathscr{C}}^2$$

$$= \langle A_0^* A_0 a, a \rangle_{\mathscr{C}}.$$

Thus (5-18) holds, and C is outer by Theorem A.　　　　　　　　　■

5.10　Outer Functions on the Half-Plane

Let S be the canonical shift operator on $H_{\mathscr{C}}^2(\Pi)$, that is, S is multiplication by $(z - i)/(z + i)$. For each $t \geq 0$, define V_t on $H_{\mathscr{C}}^2(\Pi)$ by

$$V_t: f(z) \to e^{itz}f(z).$$

The identity

$$\frac{z - i}{z + i} = 1 - 2 \int_0^\infty e^{-t}e^{itz} \, dt \tag{5-24}$$

holds for each $z \in \Pi$. We show that it also holds in an operator theoretic sense.

THEOREM A.　*We have*

$$S = I - 2 \int_0^\infty e^{-t}V_t \, dt, \tag{5-25}$$

where the integral is taken in the weak sense defined in 4.5.

Proof.　It is enough to show that

$$\langle Sf, g \rangle_2 = \langle f, g \rangle_2 - 2 \int_0^\infty e^{-t}\langle V_t f, g \rangle_2 \, dt \tag{5-26}$$

for all $f \in H_{\mathscr{C}}^2(\Pi)$ and all g in some set whose linear span is dense in $H_{\mathscr{C}}^2(\Pi)$. Choose g of the form

$$g(z) = \frac{1}{2\pi i} \frac{c}{\bar{w} - z}, \qquad z \in \Pi, \tag{5-27}$$

where $w \in \Pi$ and $c \in \mathscr{C}$. In this case, (5-26) reduces to (5-24) with $z = w$, and the result follows.　　　　　　　　　■

THEOREM B.　*The closure in the weak operator topology of the linear span of* $\{V_t\}_{t \geq 0}$ *contains* S.

Proof. By Theorem A,

$$\left\| S - I + 2 \int_0^a e^{-t} V_t \, dt \right\| = \left\| 2 \int_a^\infty e^{-t} V_t \, dt \right\| \leq 2 \int_a^\infty e^{-t} \, dt = 2e^{-a}.$$

The two integrals involving $e^{-t} V_t$ are taken in the weak sense defined in Section 4.5. It is easy to see that the Riemann sums for $\int_0^a e^{-t} V_t \, dt$ converge to the integral in the weak operator topology as well, and so the result follows. ∎

THEOREM C.   *Let* $C \in N_{\mathscr{B}(\mathscr{C})}^+(\Pi)$. *For C to be outer it is necessary and sufficient that*

$$\int_0^t |v(s)|_{\mathscr{C}}^2 \, ds \geq \int_0^t |u(s)|_{\mathscr{C}}^2 \, ds, \qquad t > 0, \tag{5-28}$$

whenever

$$\begin{cases} C(z)a(z) = \dfrac{1}{\sqrt{2\pi}} \displaystyle\int_0^\infty e^{isz} v(s) \, ds, & z \in \Pi, \\[4mm] A(z)a(z) = \dfrac{1}{\sqrt{2\pi}} \displaystyle\int_0^\infty e^{isz} u(s) \, ds, & z \in \Pi, \end{cases} \tag{5-29}$$

for some $A \in N_{\mathscr{B}(\mathscr{C})}^+(\Pi)$ *such that* $A(x)^* A(x) = C(x)^* C(x)$ *a.e. on* $(-\infty,\infty)$ *and some* $a \in N_{\mathscr{C}}^+(\Pi)$ *such that* $Ca, Aa \in H_{\mathscr{C}}^2(\Pi)$.

The integrals in (5-29) give the Paley-Wiener representation of the functions $Ca, Aa \in H_{\mathscr{C}}^2(\Pi)$ (see Section 4.8). Thus $u,v \in L_{\mathscr{C}}^2(0,\infty)$. Note that by the Plancherel theorem in Section 4.8, since

$$|A(x)a(x)|_{\mathscr{C}}^2 = |C(x)a(x)|_{\mathscr{C}}^2$$

a.e. on $(-\infty,\infty)$, we have

$$\int_0^\infty |v(s)|_{\mathscr{C}}^2 \, ds = \int_0^\infty |u(s)|_{\mathscr{C}}^2 \, ds. \tag{5-30}$$

In the sufficiency direction, the proof can easily be made to show more. Assume only that the condition holds when A is outer, and, for any fixed outer A, for a set of a's such that the span of the vectors $a(i)$ is dense in $\mathscr{C}$. Then C is outer.

Proof.   We begin with some preliminary remarks concerning the Paley-Wiener representation in Section 4.8. Every $f \in H_{\mathscr{C}}^2(\Pi)$ has a representation

$$f(z) = \frac{1}{\sqrt{2\pi}} \int_0^\infty e^{isz} h(s) \, ds, \qquad z \in \Pi,$$

where $h \in L^2_{\mathscr{C}}(0,\infty)$. It is easy to see that for any $t > 0$,

$$(V_t^* f)(z) = \frac{1}{\sqrt{2\pi}} \int_0^\infty e^{isz} h(s + t)\, ds, \qquad z \in \Pi.$$

Hence by the Plancherel theorem in Section 4.8,

$$\|V_t^* f\|_2^2 = \int_t^\infty |h(s)|_{\mathscr{C}}^2\, ds.$$

Assume that C is outer. Let u,v satisfy (5-29) for some A and a as in the theorem. By Section 5.7, Theorem A, $A = BC$ for some inner function B such that $M_{\mathrm{in}}(B) = M_{\mathrm{out}}(C)$. Applying 3.11, Lemma A, to the operator multiplication by B on $H^2_{\mathscr{C}}(\Pi)$, we obtain

$$\|V_t^* Ca\|_2^2 \leq \|V_t^* BCa\|_2^2 = \|V_t^* Aa\|_2^2$$

for all $t > 0$. By (5-29) and the remarks at the beginning of the proof, this yields

$$\int_t^\infty |v(s)|_{\mathscr{C}}^2\, ds \leq \int_t^\infty |u(s)|_{\mathscr{C}}^2\, ds, \qquad t > 0.$$

Then (5-28) follows from (5-30).

Conversely, assume that (5-28) holds whenever u,v are related as in the theorem. By 5.7, $C = BA$, where A is outer, B is inner, $M_{\mathrm{in}}(B) = M_{\mathrm{out}}(A)$, and $A(x)^* A(x) = C(x)^* C(x)$ a.e. on $(-\infty,\infty)$. For this choice of A and any a as in the theorem, there exist $u,v \in L^2_{\mathscr{C}}(0,\infty)$ satisfying (5-29), and then (5-28) holds by assumption. By what we proved above with the roles of A and C interchanged (now A is outer), equality holds in (5-28). Arguing as in the proof of necessity, we obtain

$$\|V_t^* Aa\| = \|V_t^* BAa\|_2, \qquad t \geq 0. \tag{5-31}$$

We now apply Section 3.11, Lemma B, to the operator $T(B)$ of multiplication by B on $H^2_{\mathscr{C}}(\Pi)$. In 3.11, Lemma B, choose $\{V_j\}_{j \in J}$ to be the family $\{V_t\}_{t \geq 0}$, and let $\{g_k\}_{k \in K}$ be the set of functions Aa with a as in the theorem. Notice that (3-21) holds by (5-31). By Theorem B above, the hypothesis (i) in Section 3.11, Lemma B, is satisfied. Hypothesis (ii) in 3.11, Lemma B, requires that the vectors $A(i)a(i)$ span a dense set in $M_{\mathrm{in}}(B)$, and this holds by 5.4. Hence by 3.11, Lemma B, $T(B)$ is an S-constant inner operator, and hence B is a constant inner function. Since $C = BA$, where A is outer and $M_{\mathrm{in}}(B) = M_{\mathrm{out}}(A)$, C is outer. ∎

Examples and Addenda

1. Let $A \in H^\infty_{\mathscr{B}(\mathscr{C})}(\Omega)$, $\Omega = D$ or Π, and assume that the nontangential boundary values of A are partial isometries with constant initial space M σ-a.e. on Γ or a.e. on $(-\infty, \infty)$, depending on the case. Then A is an inner function with $M_{\mathrm{in}}(A) = M$. This result is a converse to Theorem C in 5.3.

2.   If $F \in N^+_{\mathcal{B}(\mathcal{C})}(\Omega)$, $\Omega = D$ or Π, then the following are equivalent:

(i) F is an outer function;
(ii) the inequality

$$G(z)^*G(z) \le F(z)^*F(z), \qquad z \in \Omega,$$

holds whenever $G \in N^+_{\mathcal{B}(\mathcal{C})}(\Omega)$ and $G(e^{i\theta})^*G(e^{i\theta}) \le F(e^{i\theta})^*F(e^{i\theta})$ σ-a.e. on Γ or $G(x)^*G(x) \le F(x)^*F(x)$ a.e. on $(-\infty,\infty)$, depending on the case.

3.   The results sketched here are adapted from Rosenblum [1980]. Let $\mathcal{H}_1, \mathcal{H}_2$ be Hilbert spaces. Let $A, B \in \mathcal{B}(\mathcal{H}_1, \mathcal{H}_2)$ satisfy $AS = RA$, $BS = RB$, where S is an isometry in $\mathcal{B}(\mathcal{H}_1)$ and $R \in \mathcal{B}(\mathcal{H}_2)$.

THEOREM.   *There exists an operator C in $\mathcal{B}(\mathcal{H}_1)$ such that $B = AC$ and $CS = SC$ if and only if*

$$BB^* \le \gamma^2 AA^*$$

for some real constant $\gamma \ge 0$. In this case we can choose C so that in addition $\|C\| \le \gamma$.

Proof.   Necessity is clear. To prove the sufficiency part of the theorem, apply Theorem 1.14 with the quantities $\{\mathcal{H}, \mathcal{K}, S, T, R, M, M'\}$ in 1.14 chosen as $\{\mathcal{H}_1, \mathcal{H}_1, S, \gamma^2 I, S, (A^*\mathcal{H}_2)^-, (B^*\mathcal{H}_2)^-\}$. Choose the operator X in 1.14 so that $X^*: A^*h \to B^*h$ for every $h \in \mathcal{H}_2$. It is not hard to see that the hypotheses of 1.14 are satisfied. The operator Y produced by 1.14 is the required operator C.   ∎

Example 1 (R. B. Leech).   Let $A, B \in H^\infty_{\mathcal{B}(\mathcal{C})}(D)$, and let $\gamma \ge 0$ be given. Then the following are equivalent:

(i) $B = AC$ for some $C \in H^\infty_{\mathcal{B}(\mathcal{C})}(D)$ such that $\|C\|_\infty \le \gamma$;
(ii) for all finite sequences $\{z_j\} \subseteq D$ and $\{c_j\} \subseteq \mathcal{C}$,

$$\sum_{j,k} \frac{\langle A(z_k)A(z_j)^*c_j, c_k \rangle_{\mathcal{C}}}{1 - \bar{z}_j z_k} \le \gamma^2 \sum_{j,k} \frac{\langle B(z_k)B(z_j)^*c_j, c_k \rangle_{\mathcal{C}}}{1 - \bar{z}_j z_k}.$$

Example 2.   Fix $\{a_j\}_1^\infty \subseteq H^\infty(D)$ and $b \in H^\infty(D)$ such that $\sup_{z \in D} \sum_1^\infty |a_j(z)|^2$ is finite. Let $\gamma \ge 0$ be given. Then the following are equivalent:

(i) there exist functions $\{c_j\}_1^\infty \subseteq H^\infty(D)$ such that $\sup_{z \in D} \sum_1^\infty |c_j(z)|^2 \le \gamma^2$ and $b = \sum_1^\infty a_j c_j$;
(ii) $BB^* \le \gamma^2 \sum_1^\infty A_j A_j^*$, where $\{A_j\}_1^\infty$, B are the operators of multiplication by $\{a_j\}_1^\infty$, b on $H^2(D)$.

Notes

Inner-outer factorization theorems for operator valued functions were obtained by Helson and Lowdenslager [1958], [1961], Lax [1959], [1961], and Masani [1959], [1962]. See also Helson [1964] and Sz.-Nagy and Foias [1970].

In the scalar case the inner-outer factorization theory yields at the same time the complete multiplicative structure of, say, a bounded holomorphic function on the unit disk. Our treatment does not address the problem of finding the complete multiplicative structure of operator valued functions. The paper of Potapov [1955] solves this problem for matrix valued functions. There is an extensive literature devoted to extensions of Potapov's results to operator valued functions. This work is intimately connected with the theory of characteristic operator functions and triangular models for Hilbert space operators. Different perspectives on this field, which include references to the most important literature, may be found in de Branges [1965], Brodskiĭ [1969], Ginzburg [1967a,b], and Sz.-Nagy and Foias [1970].

Sections 5.9 and 5.10 follow Rosenblum and Rovnyak [1971]; Theorem C in 5.10 is due to Robinson [1962a] in the scalar case; for the significance of this result in engineering applications, see Dym and McKean [1972], pp. 175–176, and Robinson [1962b].

Chapter 6

Factorization of Nonnegative Operator Valued Functions

6.1 Introduction

If $Q(x) = Q_0 + Q_1 x + \cdots + Q_n x^n$ is a polynomial with operator coefficients, then $P(x) = Q(x)*Q(x)$ is a polynomial such that $P(x) \geq 0$ for all real x. We shall see in Section 6.7 that, conversely, every polynomial $P(x)$ with operator coefficients such that $P(x) \geq 0$ for all real x has this form.

More generally, in Chapter 6 we study the operator analogue of Szegö's problem. This is interpreted as the problem of giving conditions on an operator valued function $F(\cdot)$ on the circle Γ or line R which imply that

$$F(\cdot) = G(\cdot)*G(\cdot),$$

where $G(\cdot)$ is the boundary function of a holomorphic operator valued function of class N^+ on the unit disk D or upper half-plane Π, respectively. Our results are inspired principally by three theorems in classical function theory, due to Fejér and Riesz, Ahiezer [1948], and Szegö [1921].

FEJÉR-RIESZ THEOREM. *Any trigonometric polynomial $f(e^{i\theta}) = \sum_{-n}^{n} a_j e^{ij\theta}$ that is nonnegative on the unit circle Γ has the form $f(e^{i\theta}) = |g(e^{i\theta})|^2$, where $g(e^{i\theta}) = \sum_{0}^{n} b_j e^{ij\theta}$ is an analytic trigonometric polynomial such that $g(z) = \sum_{0}^{n} b_j z^j$ has no zeros on the disk D.*

An elementary proof can be based on the fundamental theorem of algebra. See Riesz and Sz.-Nagy [1955], pp. 117–118.

AHIEZER'S THEOREM. *Let $f(z)$ be an entire function of exponential type τ that is nonnegative on the real axis and satisfies*

$$\int_{-\infty}^{\infty} \frac{\log^+ f(x)}{1 + x^2} dx < \infty.$$

Then there exists an entire function $g(z)$ of exponential type $\tau/2$ having no zeros for $y > 0$ such that $f(x) = |g(x)|^2$ on the real axis.

Ahiezer's theorem is a generalization of the Fejér-Riesz theorem (Boas [1954], p. 125).

SZEGÖ'S THEOREM.   *Let $f(e^{i\theta})$ be a nonnegative function in $L^1(\sigma)$. For the existence of a function $g(z)$ in $H^2(D)$ having no zeros on D, such that $f(e^{i\theta}) = |g(e^{i\theta})|^2$ σ-a.e. on Γ, it is necessary and sufficient that*

$$\int_{\Gamma} \log f(e^{i\theta})\, d\sigma > -\infty.$$

We obtain extensions of these results to operator valued functions. The operator versions of the Fejér-Riesz and Ahiezer theorems follow as special cases of a general factorization theorem for pseudomeromorphic functions (Theorem 6.5). We also prove a generalization of Kreǐn's theorem (Appendix, Section 5) for operator valued functions.

Notation is as in Chapters 4 and 5. Thus $\mathscr{C}$ denotes a separable Hilbert space. By S we always mean the canonical shift operator defined on $H^2_\mathscr{C}(D)$ or $H^2_\mathscr{C}(\Pi)$ as in Section 5.2. Equivalently, we may view S as acting on boundary functions, so that either

$$S: f(e^{i\theta}) \to e^{i\theta} f(e^{i\theta}) \quad \text{on } H^2_\mathscr{C}(\Gamma)$$

or

$$S: f(x) \to \frac{x-i}{x+i} f(x) \quad \text{on } H^2_\mathscr{C}(R),$$

depending on the case.

6.2   Toeplitz Operators and the Factorization of Nonnegative Operator Valued Functions

Consider the disk case. We describe the class of S-Toeplitz operators on $H^2_\mathscr{C}(\Gamma)$ and relate the factorization properties of these operators to the problem at hand. Let P be the projection of $L^2_\mathscr{C}(\sigma)$ on $H^2_\mathscr{C}(\Gamma)$.

THEOREM A.   *A bounded linear operator T on $H^2_\mathscr{C}(\Gamma)$ is S-Toeplitz in the sense of Section 3.1 if and only if $T = T(W)$, where*

$$T(W): f \to PWf, \qquad f \in H^2_\mathscr{C}(\Gamma), \tag{6-1}$$

for some $W \in L^\infty_{\mathscr{B}(\mathscr{C})}(\sigma)$. In this case, $\|T\| = \|W\|_\infty$.

When $W \in H^\infty_{\mathscr{B}(\mathscr{C})}(\Gamma)$, $T(W)$ is S-analytic, and (6-1) may be written

$$T(W): f \to Wf, \qquad f \in H^2_\mathscr{C}(\Gamma). \tag{6-2}$$

Conversely, by 1.15, Theorem B, every S-analytic operator has this form.

THEOREM B.   *Let $W \in L^\infty_{\mathscr{B}(\mathscr{C})}(\sigma)$ and $A \in H^\infty_{\mathscr{B}(\mathscr{C})}(\Gamma)$. Then:*

(i) $T(W) \geq 0$ *if and only if* $W(e^{i\theta}) \geq 0$ *σ-a.e. on Γ;*
(ii) $T(W) = T(A)^*T(A)$ *if and only if* $W(e^{i\theta}) = A(e^{i\theta})^* A(e^{i\theta})$ *σ-a.e. on Γ.*

LEMMA.   *A bounded linear operator L on $L^2_\mathscr{C}(\sigma)$ commutes with the operator*

$$U: f(e^{i\theta}) \to e^{i\theta} f(e^{i\theta}) \tag{6-3}$$

on $L^2_{\mathscr{C}}(\sigma)$ if and only if

$$L: f \to Wf \tag{6-4}$$

for some $W \in L^\infty_{\mathscr{B}(\mathscr{C})}(\sigma)$. In this case, $\|W\|_\infty = \|L\|$.

Proof of lemma.   Clearly any operator of the form (6-4) commutes with U and $\|L\| \le \|W\|_\infty$.

Conversely, suppose that $LU = UL$. Then $LU^j = U^j L$ for all $j = 0, \pm 1, \pm 2, \ldots$. Hence

$$L(\phi f) = \phi \cdot (Lf), \qquad f \in L^2_{\mathscr{C}}(\sigma), \tag{6-5}$$

for any trigonometric polynomial ϕ with scalar coefficients. By a routine approximation argument, (6-5) holds for all continuous complex valued functions ϕ on Γ.

Let $\mathscr{E}$ be a countable dense set in $\mathscr{C}$. For each $c \in \mathscr{C}$, let $g_c(e^{i\theta})$ be a representative in the coset Lc. For any fixed $a,b \in \mathscr{E}$,

$$\lim_{w \to e^{i\theta}} \int_\Gamma P(w,e^{it})\langle g_a(e^{it}),b\rangle_{\mathscr{C}}\, d\sigma = \langle g_a(e^{i\theta}),b\rangle_{\mathscr{C}} \tag{6-6}$$

nontangentially σ-a.e. on Γ by Fatou's theorem. Since $\mathscr{E}$ is countable, we can choose a σ-null set $N \subseteq \Gamma$ such that (6-6) holds for all $a,b \in \mathscr{E}$ and $e^{i\theta} \in \Gamma \backslash N$.

Fix $e^{i\theta} \in \Gamma \backslash N$. Define $s_0(e^{i\theta}; \cdot, \cdot)$ on $\mathscr{E} \times \mathscr{E}$ by

$$s_0(e^{i\theta}; a,b) = \langle g_a(e^{i\theta}),b\rangle_{\mathscr{C}}, \qquad a,b \in \mathscr{E}.$$

Define $\phi_w(e^{it}) = P(w,e^{it})^{1/2}$ for $w \in D$, $e^{it} \in \Gamma$. By (6-5),

$$\int_\Gamma P(w,e^{it})\langle g_a(e^{it}),b\rangle_{\mathscr{C}}\, d\sigma = \langle \phi_w(La),\phi_w b\rangle_2 = \langle L(\phi_w a),\phi_w b\rangle_2,$$

so by (6-6),

$$s_0(e^{i\theta}; a,b) = \lim_{w \to e^{i\theta}} \langle L(\phi_w a),\phi_w b\rangle_2 \tag{6-7}$$

nontangentially for all $a,b \in \mathscr{E}$. It is easy to see that the limit on the right of (6-7) exists for all $a,b \in \mathscr{C}$ and defines a bounded sesquilinear form $s(e^{i\theta}; \cdot, \cdot)$ on $\mathscr{C} \times \mathscr{C}$ that extends $s_0(e^{i\theta}; \cdot, \cdot)$ and satisfies $\|s\| \le \|L\|$. Hence there is an operator $W(e^{i\theta}) \in \mathscr{B}(\mathscr{C})$ such that $|W(e^{i\theta})|_{\mathscr{B}(\mathscr{C})} \le \|L\|$ and

$$s(e^{i\theta}; a,b) = \langle W(e^{i\theta})a,b\rangle_{\mathscr{C}}, \qquad a,b \in \mathscr{C}.$$

Now consider $W(e^{i\theta})$ as a function of $e^{i\theta}$. By construction, $W \in L^\infty_{\mathscr{B}(\mathscr{C})}(\sigma)$ and $\|W\|_\infty \le \|L\|$. For $c \in \mathscr{E}$, $L: c \to W(e^{i\theta})c$. In a straightforward way we obtain (6-4), and the result follows.   ∎

Proof of Theorem A.   Let $T(W)$ be defined by (6-1) for some $W \in L^\infty_{\mathscr{B}(\mathscr{C})}(\sigma)$. Clearly $\|T(W)\| \le \|W\|_\infty$. For any $f,g \in H^2_{\mathscr{C}}(\Gamma)$,

$$\langle T(W)Sf,Sg\rangle_2 = \int_\Gamma \langle W(e^{i\theta})e^{i\theta}f(e^{i\theta}),e^{i\theta}g(e^{i\theta})\rangle_{\mathscr{C}}\, d\sigma = \langle T(W)f,g\rangle_2.$$

Hence $S^*T(W)S = T(W)$, that is, $T(W)$ is S-Toeplitz. Sufficiency follows.

Conversely, let T be an S-Toeplitz operator on $H^2_\mathscr{C}(\Gamma)$. Set $\mathscr{H}_n = U^{-n}H^2_\mathscr{C}(\Gamma)$, $n = 0,1,2,\ldots$, where U is the operator on $L^2_\mathscr{C}(\sigma)$ defined by (6-3). Then $\bigcup_0^\infty \mathscr{H}_n$ is dense in $L^2_\mathscr{C}(\sigma)$ and

$$H^2_\mathscr{C}(\Gamma) = \mathscr{H}_0 \subseteq \mathscr{H}_1 \subseteq \mathscr{H}_2 \subseteq \cdots \subseteq L^2_\mathscr{C}(\sigma).$$

For each $n \geq 0$, set

$$s_n(f,g) = \langle TU^n f, U^n g \rangle_2, \qquad f,g \in \mathscr{H}_n.$$

Then $s_n(\cdot,\cdot)$ is a bounded sesquilinear form on $\mathscr{H}_n$ with $\|s_n\| \leq \|T\|$. Since $S^*TS = T$ by assumption and $S = U|\mathscr{H}_0$, for any $f,g \in \mathscr{H}_n$,

$$s_{n+1}(f,g) = \langle TU^{n+1}f, U^{n+1}g \rangle_2 = \langle TSU^n f, SU^n g \rangle_2$$
$$= \langle TU^n f, U^n g \rangle_2 = s_n(f,g).$$

Hence there is a bounded sesquilinear form $s(\cdot,\cdot)$ on $L^2_\mathscr{C}(\sigma)$ that extends each $s_n(\cdot,\cdot)$ and satisfies $\|s\| \leq \|T\|$. Let L be the unique operator on $L^2_\mathscr{C}(\sigma)$ such that

$$s(f,g) = \langle Lf,g \rangle_2, \qquad f,g \in L^2_\mathscr{C}(\sigma),$$

and $\|L\| = \|s\| \leq \|T\|$. For $f,g \in \mathscr{H}_{n+1}, n \geq 0$,

$$\langle LUf, Ug \rangle_2 = s(Uf,Ug) = s_n(Uf,Ug)$$
$$= \langle TU^n Uf, U^n Ug \rangle_2 = s_{n+1}(f,g) = s(f,g) = \langle Lf,g \rangle_2.$$

Therefore $U^*LU = L$ and $LU = UL$. By the lemma, L has the form (6-4) for some $W \in L^\infty_{\mathscr{B}(\mathscr{C})}(\sigma)$ with $\|W\|_\infty = \|L\| \leq \|T\|$. For any $f,g \in H^2_\mathscr{C}(\Gamma)$,

$$\langle Tf,g \rangle_2 = s_0(f,g) = \langle Lf,g \rangle_2$$

$$= \int_\Gamma \langle W(e^{i\theta})f(e^{i\theta}), g(e^{i\theta}) \rangle_\mathscr{C}\, d\sigma = \langle T(W)f,g \rangle_2.$$

It follows that $T = T(W)$. By construction, $\|W\|_\infty \leq \|T\|$. The reverse inequality holds automatically as in the proof of sufficiency, so the result follows. ∎

Proof of Theorem B.   (i) If $W(e^{i\theta}) \geq 0$ σ-a.e. on Γ, then

$$\langle T(W)f,f \rangle_2 = \int_\Gamma \langle W(e^{i\theta})f(e^{i\theta}), f(e^{i\theta}) \rangle_\mathscr{C}\, d\sigma \geq 0$$

for every $f \in H^2_\mathscr{C}(\Gamma)$, so $T(W) \geq 0$.

Conversely, let $T = T(W) \geq 0$. Construct L on $L^2_\mathscr{C}(\sigma)$ as in the proof of Theorem A. Then $L \geq 0$, so

$$\int_\Gamma \langle W(e^{i\theta})f(e^{i\theta}), f(e^{i\theta}) \rangle_\mathscr{C}\, d\sigma = \langle Lf,f \rangle_2 \geq 0$$

for every $f \in L^2_\mathscr{C}(\sigma)$. It follows that $W(e^{i\theta}) \geq 0$ σ-a.e. on Γ.

(ii) Since $A \in H^\infty_{\mathscr{B}(\mathscr{C})}(\Gamma)$, $T(A)$ is multiplication by A on $H^2_\mathscr{C}(\Gamma)$. Thus

$T(W) = T(A)^*T(A)$ if and only if for all $f,g \in H^2_{\mathscr{C}}(\Gamma)$,

$$\langle T(W)f,g \rangle_2 = \langle Af,Ag \rangle_2,$$

that is,

$$\int_\Gamma \langle W(e^{i\theta})f(e^{i\theta}),g(e^{i\theta}) \rangle_{\mathscr{C}} \, d\sigma = \int_\Gamma \langle A(e^{i\theta})f(e^{i\theta}),A(e^{i\theta})g(e^{i\theta}) \rangle_{\mathscr{C}} \, d\sigma,$$

or equivalently $W(e^{i\theta}) = A(e^{i\theta})^*A(e^{i\theta})$ σ-a.e. on Γ. ∎

6.3 Pseudomeromorphic Functions

The exterior of the unit circle is denoted $\tilde{D}$; the lower half-plane, $\tilde{\Pi}$; that is,

$$\tilde{D} = \{z : |z| > 1\} \qquad \text{and} \qquad \tilde{\Pi} = \{z : \operatorname{Im} z < 0\}. \tag{6-8}$$

If F is a $\mathscr{B}(\mathscr{C})$-valued function on a set $\Omega \subseteq \mathbf{C}$, the *reflection of F with respect to Γ* is the function

$$\tilde{F}(z) = F(1/\bar{z})^* \quad \text{on } \tilde{\Omega} = \{z : 1/\bar{z} \in \Omega\}. \tag{6-9}$$

The *reflection of F with respect to R* is

$$\tilde{F}(z) = F(\bar{z})^* \quad \text{on } \tilde{\Omega} = \{z : \bar{z} \in \Omega\}. \tag{6-10}$$

Whether (6-9) or (6-10) is intended will either be clear from context or indicated.

The notion of a Laurent expansion has a routine extension to $\mathscr{B}(\mathscr{C})$-valued functions. Removable singularities, essential singularities, poles, and principal parts are then defined in the usual way. We assume that removable singularities have been removed. A $\mathscr{B}(\mathscr{C})$-valued function F is meromorphic on an open set Ω in $\mathbf{C}_\infty = \mathbf{C} \cup \{\infty\}$ if it is holomorphic on Ω except for poles.

DEFINITION A. (i) *Let u,v be nonzero scalar valued functions in $N^+(\Gamma)$. A $\mathscr{B}(\mathscr{C})$-valued function F on Γ is of class $\mathscr{M}(u,v)$ if $uF, vF^* \in N^+_{\mathscr{B}(\mathscr{C})}(\Gamma)$.*

(ii) *Let u,v be nonzero scalar valued functions in $N^+(R)$. A $\mathscr{B}(\mathscr{C})$-valued function F on R is of class $\mathscr{M}(u,v)$ if $uF,vF^* \in N^+_{\mathscr{B}(\mathscr{C})}(R)$.*

Functions of class $\mathscr{M}(u,v)$ are called *pseudomeromorphic* because of the characterizations in Theorems A and B below. The class $\mathscr{M}(u,v)$ does not depend on the outer factors in u and v: F is of class $\mathscr{M}(u,v)$ if and only if it is of class $\mathscr{M}(u_i,v_i)$, where u_i,v_i are the inner factors of u,v, respectively.

Example. Let $u(e^{i\theta}) = e^{im\theta}$, $v(e^{i\theta}) = e^{in\theta}$, where m,n are nonnegative integers. Every trigonometric polynomial

$$F(e^{i\theta}) = \sum_{-m}^{n} A_j e^{ij\theta} \tag{6-11}$$

with coefficients in $\mathscr{B}(\mathscr{C})$ is of class $\mathscr{M}(u,v)$.

(i) If $F \in L^1_{\mathscr{B}(\mathscr{C})}(\sigma)$ and F is of class $\mathscr{M}(u,v)$, then F has the form (6-11).

For, by 4.7, Theorem A, $uF, vF^* \in N^+_{\mathscr{B}(\mathscr{C})}(\Gamma) \cap L^1_{\mathscr{B}(\mathscr{C})}(\Gamma) = H^1_{\mathscr{B}(\mathscr{C})}(\Gamma)$. Hence by 4.7, Theorem C, for all $j \geq 1$,

$$\int_\Gamma e^{ij\theta} u(e^{i\theta}) F(e^{i\theta})\, d\sigma = \int_\Gamma e^{ij\theta} v(e^{i\theta}) F(e^{i\theta})^*\, d\sigma = 0.$$

Therefore for $j > m$ or $j < -n$,

$$\int_\Gamma e^{ij\theta} F(e^{i\theta})\, d\sigma = 0.$$

By the Cauchy representation (4-24), F has the form (6-11).

(ii) For each $p \in (0,1)$, there is a function $F \in L^p_{\mathscr{B}(\mathscr{C})}(\sigma)$ of class $\mathscr{M}(u,v)$ and not of the form (6-11).

An example is $F(e^{i\theta}) = F_0/(e^{i\theta} - 1)$ for any nonzero $F_0 \in \mathscr{B}(\mathscr{C})$.

THEOREM A.   *Let $u_0(e^{i\theta})$, $v_0(e^{i\theta})$ be nonzero scalar valued functions in $N^+(\Gamma)$, and let $u(z), v(z)$ be the corresponding functions in $N^+(D)$. Let $F(z)$ be a $\mathscr{B}(\mathscr{C})$-valued meromorphic function on $D \cup \tilde{D}$ such that:*

(i) *the restrictions of uF and $v\tilde{F}$ to D are in $N^+_{\mathscr{B}(\mathscr{C})}(D)$;*
(ii) *$F(re^{i\theta})$ has the same strong limit $F_0(e^{i\theta})$ for $r \uparrow 1$ and $r \downarrow 1$ σ-a.e. on Γ.*

Then $F_0(e^{i\theta})$ is of class $\mathscr{M}(u_0, v_0)$. Conversely, every function of class $\mathscr{M}(u_0, v_0)$ has this form.

THEOREM B.   *Let $u_0(x), v_0(x)$ be nonzero scalar valued functions in $N^+(R)$, and let $u(z), v(z)$ be the corresponding functions in $N^+(\Pi)$. Let $F(z)$ be a $\mathscr{B}(\mathscr{C})$-valued meromorphic function on $\Pi \cup \tilde{\Pi}$ such that:*

(i) *the restrictions of uF and $v\tilde{F}$ to Π are in $N^+_{\mathscr{B}(\mathscr{C})}(\Pi)$;*
(ii) *$F(x + iy)$ has the same strong limit $F_0(x)$ for $y \downarrow 0$ and $y \uparrow 0$ a.e. on $(-\infty, \infty)$.*

Then $F_0(x)$ is of class $\mathscr{M}(u_0, v_0)$. Conversely, every function of class $\mathscr{M}(u_0, v_0)$ has this form.

DEFINITION B.   *In the situation of either Theorem A or Theorem B, we say that F is of* class $\mathscr{M}(u,v)$. *We refer to F_0 as the* boundary function *of F.*

Proof of Theorem A.   The restrictions of uF and $v\tilde{F}$ to D have boundary functions $u_0 F_0$ and $v_0 F_0^*$. These functions therefore belong to $N^+_{\mathscr{B}(\mathscr{C})}(\Gamma)$, and hence F_0 is of class $\mathscr{M}(u_0, v_0)$.

Conversely, let G_0 be any $\mathscr{B}(\mathscr{C})$-valued function of class $\mathscr{M}(u_0, v_0)$. Then $u_0 G_0, v_0 G_0^* \in N^+_{\mathscr{B}(\mathscr{C})}(\Gamma)$, and so $u_0 G_0, v_0 G_0^*$ are the boundary functions of some functions G_+, G_- in $N^+_{\mathscr{B}(\mathscr{C})}(D)$. Set

$$G(z) = \begin{cases} G_+(z)/u(z), & z \in D, \\ \tilde{G}_-(z)/\tilde{v}(z), & z \in \tilde{D}. \end{cases}$$

A routine check shows that G_0 and G are related in the required manner. ∎

Proof of Theorem B.   This follows from Theorem A by a change of variables. ∎

6.4 Analyticity Across the Boundary

Let $\Omega = D$ or Π, and let Δ be an open subset of $\partial\Omega$. Define $\tilde{\Omega} = \tilde{D}$ or $\tilde{\Pi}$ as in (6-8).

DEFINITION. *A meromorphic $\mathcal{B}(\mathscr{C})$-valued function F on $\Omega \cup \tilde{\Omega}$ is analytic across Δ if F can be defined on Δ so that when viewed as a function on $\Omega \cup \tilde{\Omega} \cup \Delta$, F is holomorphic at each point of Δ.*

A holomorphic scalar valued function f on Ω is said to have an analytic continuation across Δ if $f = g|\Omega$, where g is holomorphic on some open set G containing $\Omega \cup \Delta$. The following result generalizes a theorem of Carleman [1944], p. 40.

THEOREM. *Let F be a meromorphic $\mathcal{B}(\mathscr{C})$-valued function on $\Omega \cup \tilde{\Omega}$ of class $\mathscr{M}(u,v)$ for some nonzero functions $u,v \in N^+(\Omega)$. Assume that:*

(i) *u,v have analytic continuations across Δ;*
(ii) *if F_0 is the boundary function of F, then for each $a \in \mathscr{C}$, the scalar valued function $\langle F_0(\cdot)a,a\rangle_{\mathscr{C}}$ is integrable over every compact subset of Δ.*

Then F is analytic across Δ.

The functions in the lemma below are scalar valued.

LEMMA. *Let $f \in N^+(\Pi)$ and suppose that $\int_a^b |f(x)|^p\,dx < \infty$, where $-\infty < a < b < \infty$ and $0 < p < \infty$. Then*

$$\lim_{y\downarrow 0} \int_c^d |f(x) - f(x + iy)|^p\,dx = 0$$

for every closed subinterval $[c,d]$ of (a,b).

Proof of lemma. Choose q such that $pq > 1$. Let g be an outer function such that $|g(x)| = 1$ on (a,b) and $|g(x)| = (|x| + 1)^q|f(x)|$ otherwise. Then $h = f/g \in N^+(\Pi)$ and $\int_{-\infty}^{\infty} |h(x)|^p\,dx < \infty$. Therefore $h \in H^p(\Pi)$ and

$$\lim_{y\downarrow 0} \int_{-\infty}^{\infty} |h(x) - h(x + iy)|^p\,dx = 0$$

(see Krylov [1939]). The function g has an analytic continuation across (a,b), and so $\lim_{y\downarrow 0} g(x + iy) = g(x)$ uniformly on every closed subinterval $[c,d]$ of (a,b). In view of the elementary inequality

$$|u + v|^p \le (\max(2|u|,2|v|))^p \le 2^p(|u|^p + |v|^p),$$

this is sufficient to imply the lemma. ∎

Proof of theorem. We give the proof for the case $\Omega = \Pi$. Then the other case follows by a change of variables.

We may assume that $\Delta = (\alpha,\beta)$, $-\infty < \alpha < \beta < \infty$, and that the analytic continuations of u and v across Δ have no zeros on Δ. For if there are zeros, we can contract the interval slightly and reduce to the case of a finite number of zeros; dividing out factors to remove these zeros does not change $\mathcal{M}(u,v)$ because the factors are outer functions.

Consider an interval $[c,d] \subseteq (\alpha,\beta)$ such that $F(x + iy)$ has a strong limit for $x = c,d$ as $|y| \to 0$. We can choose such an interval with c arbitrarily near α and d arbitrarily near β. Choose a rectangle $Q = (c,d) \times (-\delta,\delta)$, where $\delta > 0$ is small enough that the analytic continuations of u,v across Δ are defined and nonvanishing on $\bar{Q}$. Define

$$G(z) = \frac{1}{2\pi i} \int\limits_{\partial Q} \frac{F(t)}{t - z} dt, \qquad z \in Q.$$

Our assumptions imply that F is sufficiently regular on ∂Q for the integral to exist in the weak sense: $F(x + iy)$ remains bounded for $x = c,d, |y| \to 0$, by the uniform boundedness principle. The function $G(z)$ is holomorphic on Q. To complete the proof, we show that F coincides with G on $Q \cap (\Pi \cup \tilde{\Pi})$. By considering the scalar valued functions $\langle F(\cdot)a,a \rangle_\mathscr{C}$ and $\langle G(\cdot)a,a \rangle_\mathscr{C}$ for arbitrary $a \in \mathscr{C}$, we can assume without loss of generality that we are in the scalar case, that is, F and G are themselves scalar valued functions.

Set $Q(\varepsilon+) = (c,d) \times (\varepsilon,\delta)$ for any $\varepsilon \in (0,\delta)$. By Cauchy's theorem,

$$F(z) = \frac{1}{2\pi i} \int\limits_{\partial Q(\varepsilon+)} \frac{F(t)}{t - z} dt, \qquad z \in Q(\varepsilon+).$$

We show that

$$\lim_{y \downarrow 0} \int\limits_c^d |F(x + iy) - F(x)| \, dx = 0. \tag{6-12}$$

By assumption, F is of class $\mathcal{M}(u,v)$, so $uF \in N^+(\Pi)$. By the lemma, (6-12) holds with F replaced by uF. The assumptions on u imply that we can drop the factor u, and (6-12) follows. Letting $\varepsilon \downarrow 0$, we obtain

$$F(z) = \frac{1}{2\pi i} \int\limits_{\partial Q+} \frac{F(t)}{t - z} dt, \qquad z \in Q+,$$

where $Q+ = Q \cap \Pi$ and the boundary function of F is used on the lower edge. Combining this formula with an analogous formula for $Q- = Q \cap \tilde{\Pi}$, we obtain

$$F(z) = \frac{1}{2\pi i} \int\limits_{\partial Q-} \frac{F(t)}{t - z} dt, \qquad z \in Q \cap (\Pi \cup \tilde{\Pi}).$$

Thus $F = G$ on $Q \cap (\Pi \cup \tilde{\Pi})$, and the result follows. ∎

We can now state and prove one of the main results of Chapter 6.

6.5   Theorem. Factorization of Pseudomeromorphic Functions

Let v be any nonzero scalar valued function in $N^+(\Gamma)$ or $N^+(R)$. If F is any nonnegative $\mathscr{B}(\mathscr{C})$-valued function of class $\mathscr{M}(v,v)$ on Γ or R, then

$$F = G^*G \qquad\qquad (6\text{-}13)$$

σ-a.e. on Γ or a.e. on R, where G is an outer function of class $\mathscr{M}(1,v)$ on Γ or R, respectively.

The factorization is essentially unique by Section 5.8, Theorem B.

Proof.   We give the proof in the circle case. The other case then follows by a change of variables. Since $\mathscr{M}(v,v)$ does not depend on the outer factor in v, we may assume that v is inner.

Since F is of class $\mathscr{M}(v,v)$, $vF \in N^+_{\mathscr{B}(\mathscr{C})}(\Gamma)$. By Section 4.3, Theorem C, there is a bounded scalar valued outer function ϕ on Γ such that $\phi vF \in H^\infty_{\mathscr{B}(\mathscr{C})}(\Gamma)$. Since $|v| = 1$ σ-a.e. on Γ, the function $W = \bar{\phi}F\phi$ belongs to $L^\infty_{\mathscr{B}(\mathscr{C})}(\sigma)$. Let P be the projection of $L^2_\mathscr{C}(\sigma)$ on $H^2_\mathscr{C}(\Gamma)$, and define the Toeplitz operator $T(W)$ by (6-1). By Section 6.2, Theorem B, $T(W) \geq 0$, and thus Theorem 3.4 is applicable. We show that condition (iii) of 3.4 is satisfied. For each $c \in \mathscr{C}$ and $n = 0,1,2,\ldots$, set

$$I_n(c) = \sup\{|\langle T(W)c,S^nf \rangle_2|: f \in H^2_\mathscr{C}(\Gamma), \langle T(W)f,f \rangle_2 = 1\}.$$

Here c is viewed as a constant function in $H^2_\mathscr{C}(\Gamma)$. Set

$$\chi(e^{i\theta}) = e^{i\theta}, \qquad e^{i\theta} \in \Gamma.$$

For any $f \in H^2_\mathscr{C}(\Gamma)$,

$$|\langle T(W)c,S^nf \rangle_2| = \left|\int_\Gamma \langle Wc,\chi^nf \rangle_\mathscr{C}\, d\sigma\right|$$

$$= \left|\int_\Gamma \langle (\phi/\bar{\phi})vc,\chi^nP\phi^2vFf \rangle_\mathscr{C}\, d\sigma\right|$$

$$= \left|\int_\Gamma \langle P\chi^{-n}P(\phi/\bar{\phi})vc,v(F^{1/2}\phi)^2f \rangle_\mathscr{C}\, d\sigma\right|$$

$$= \left|\int_\Gamma \langle F^{1/2}\bar{\phi}P\chi^{-n}P(\phi/\bar{\phi})vc,vF^{1/2}\phi f \rangle_\mathscr{C}\, d\sigma\right|$$

$$\leq \left(\int_\Gamma |F^{1/2}\bar{\phi}P\chi^{-n}P(\phi/\bar{\phi})vc|^2_\mathscr{C}\, d\sigma\right)^{1/2} \langle T(W)f,f \rangle_2.$$

By the choice of ϕ, $F^{1/2}\phi$ is bounded. Hence

$$I_n(c) \leq \text{const.}\left(\int_\Gamma |P\chi^{-n}P(\phi/\bar{\phi})vc|^2_\mathscr{C}\, d\sigma\right)^{1/2}$$

$$= \text{const.}\|S^{*n}g_c\|_2,$$

where $g_c = P(\phi/\bar{\phi})vc$. Since S is a shift operator, $I_n(c) \to 0$ as $n \to \infty$. Thus condition (iii) of 3.4 is satisfied. By Theorem 3.4 and Section 6.2, there is an outer function A such that $W = A^*A$ σ-a.e. on Γ. Therefore

$$F = W/(\bar{\phi}\phi) = A^*A/(\bar{\phi}\phi) = G^*G$$

σ-a.e. on Γ, where $G = A/\phi$ is outer.

We show that G is of class $\mathcal{M}(1,v)$. Consider the function $C = vA^*(\phi/\bar{\phi})$ in $L^\infty_{\mathcal{B}(\mathcal{C})}(\sigma)$. By the choice of ϕ, $CA = \phi^2vF \in H^\infty_{\mathcal{B}(\mathcal{C})}(\Gamma)$. Thus

$$C(AH^2_{\mathcal{C}}(\Gamma))^- \subseteq H^2_{\mathcal{C}}(\Gamma).$$

Since A is outer, $(AH^2_{\mathcal{C}}(\Gamma))^-$ reduces S. Hence if g is in $H^2_{\mathcal{C}}(\Gamma)$ and orthogonal to $(AH^2_{\mathcal{C}}(\Gamma))^-$, so is $S^jg, j = 0,1,2,\ldots$. Then for any $u \in H^2_{\mathcal{C}}(\Gamma)$ and $j \geq 0$,

$$\int_\Gamma \langle A^*g, \chi^{-j}u \rangle_{\mathcal{C}}\, d\sigma = \int_\Gamma \langle \chi^jg, Au \rangle_{\mathcal{C}}\, d\sigma = \langle S^jg, Au \rangle_2 = 0.$$

Therefore $A^*g = 0$ σ-a.e. on Γ, and so $Cg = 0$ σ-a.e. on Γ. It follows that

$$CH^2_{\mathcal{C}}(\Gamma) \subseteq H^2_{\mathcal{C}}(\Gamma),$$

and hence $C \in H^\infty_{\mathcal{B}(\mathcal{C})}(\Gamma)$. Therefore $\phi vG^* = C \in H^\infty_{\mathcal{B}(\mathcal{C})}(\Gamma)$, and so $vG^* \in N^+_{\mathcal{B}(\mathcal{C})}(\Gamma)$. Trivially $1 \cdot G \in N^+_{\mathcal{B}(\mathcal{C})}(\Gamma)$, and hence G is of class $\mathcal{M}(1,v)$. This completes the proof.   ∎

6.6   Operator Fejér-Riesz Theorem

Let $F(e^{i\theta}) = \sum^n_{-n} A_je^{ij\theta}$ be a trigonometric polynomial with coefficients in $\mathcal{B}(\mathcal{C})$ such that $F(e^{i\theta}) \geq 0$ on Γ. Then

$$F(e^{i\theta}) = G(e^{i\theta})^*G(e^{i\theta}), \qquad e^{i\theta} \in \Gamma, \tag{6-14}$$

where $G(e^{i\theta})$ is an outer function of the form $G(e^{i\theta}) = \sum^n_0 B_je^{ij\theta}$ with coefficients in $\mathcal{B}(\mathcal{C})$.

Proof. In view of the example in Section 6.3, this follows as a special case of Theorem 6.5.   ∎

6.7   Theorem. Polynomials That Are Nonnegative on a Line

Let $P(x) = \sum^{2n}_0 P_jx^j$ be a polynomial with coefficients in $\mathcal{B}(\mathcal{C})$ such that $P(x) \geq 0$ for all real x. Then

$$P(x) = Q(x)^*Q(x), \qquad x \in R, \tag{6-15}$$

where $Q(x)$ is an outer function of the form $Q(x) = \sum^n_0 Q_jx^j$.

Proof. Set $v(x) = (x - i)^n/(x + i)^n$ and $F(x) = (x^2 + 1)^{-n}P(x)$. Then $v(x)F(x)$ and $v(x)F(x)^*$ are the boundary functions of functions that are bounded and holomorphic on Π. Hence F is of class $\mathcal{M}(v,v)$, so by 6.5, $F(x) = G(x)^*G(x)$ a.e. on $(-\infty,\infty)$, where $G(x)$ is an outer function of class $\mathcal{M}(1,v)$. Consider the associated meromorphic function $G(z)$ as in Section 6.3, Theorem B. Since $G(x)$ and $v(x)G(x)^*$ are bounded functions in $N^+_{\mathscr{B}(\mathscr{C})}(R)$, $G(z)$ and $v(z)\tilde{G}(z)$ are bounded on Π (Examples and Addenda to Chapter 4, no. 3). Hence

$$|(z + i)^n G(z)| \leq M(|z| + 1)^n, \qquad z \in \Pi \cup \tilde{\Pi},$$

for some constant $M > 0$. By Section 6.4, $(z + i)^n G(z)$ is an entire function. Therefore by the Cauchy estimates, $Q(z) = (z + i)^n G(z)$ is a polynomial of degree at most n. Since $G(x)$ is outer as a function on $(-\infty,\infty)$, so is $Q(x)$. By construction, (6-15) holds. ∎

6.8   Rational Functions

A $\mathscr{B}(\mathscr{C})$-valued function $F(z)$ is called *rational* if it is meromorphic on $\mathbf{C}_\infty = \mathbf{C} \cup \{\infty\}$. It is not hard to see that $F(z)$ is rational if and only if $F(z) = P(z)/q(z)$, where $P(z)$ is a $\mathscr{B}(\mathscr{C})$-valued polynomial and $q(z)$ is a scalar valued polynomial.

THEOREM.   *Let $F(z)$ be a $\mathscr{B}(\mathscr{C})$-valued rational function that is either nonnegative at all points $e^{i\theta} \in \Gamma$ that are not poles, or nonnegative at all points $x \in R$ that are not poles. In the circle case,*

$$F(z) = \tilde{G}(z)G(z),$$

where $G(z)$ is a $\mathscr{B}(\mathscr{C})$-valued rational function that is holomorphic on D and whose restriction to D is outer. In the line case there is a similar factorization with respect to the half-plane Π.

The tilde notation is as in Section 6.3. Thus $\tilde{G}(z) = G(1/\bar{z})^*$ or $\tilde{G}(z) = G(\bar{z})^*$, depending on the case.

Proof.   In the disk case, choose a scalar polynomial $q(z)$ whose restriction to D is outer such that $P(z) = \tilde{q}(z)q(z)F(z)$ is a polynomial in z and $1/z$. Then $P(e^{i\theta}) \geq 0$ on Γ, so by 6.6,

$$P(e^{i\theta}) = Q(e^{i\theta})^* Q(e^{i\theta}),$$

where $Q(z)$ is a polynomial whose restriction to D is outer. The required factorization is obtained with $G(z) = Q(z)/q(z)$.

In the half-plane case, choose a scalar polynomial $q(z)$ whose restriction to D is outer such that $P(z) = \tilde{q}(z)q(z)F(z)$ is a polynomial in z. Argue as above using 6.7 instead of 6.6. ∎

6.9   Functions of Bounded Type; Mean Type

The notion of mean type for scalar valued functions in $N(\Pi)$ is defined in the Appendix, Section 6. We now extend this notion to functions F in $N_{\mathscr{B}(\mathscr{C})}(\Pi)$. For each $c \in \mathscr{C}$, define F_c by

$$F_c(z) = \langle F(z)c,c \rangle_{\mathscr{C}}, \qquad z \in \Pi.$$

DEFINITION.   *The* mean type *of a function F in* $N_{\mathscr{B}(\mathscr{C})}(\Pi)$ *is the number* $\tau = \sup_{c \in \mathscr{C}} \tau_c$, *where* τ_c *is the mean type of* F_c *for any* $c \in \mathscr{C}$.

THEOREM.   *The mean type* τ *of any F in* $N_{\mathscr{B}(\mathscr{C})}(\Pi)$ *satisfies* $-\infty \leq \tau < \infty$, *with* $\tau = -\infty$ *only if* $F \equiv 0$.

Proof.   Let τ_c be the mean type of F_c for any $c \in \mathscr{C}$. By Theorem E in Section 6 of the Appendix,

$$\tau_c = \limsup_{y \to \infty} y^{-1} \log |F_c(iy)| \leq \limsup_{y \to \infty} y^{-1} \log |F(iy)|_{\mathscr{B}(\mathscr{C})}.$$

By Section 4.3, Theorem A, there is a scalar valued holomorphic function u such that $0 < |u| \leq 1$ and $|uF|_{\mathscr{B}(\mathscr{C})} \leq 1$ on Π. Then

$$
\begin{aligned}
\limsup_{y \to \infty} & y^{-1} \log |F(iy)|_{\mathscr{B}(\mathscr{C})} \\
&= \limsup_{y \to \infty} \left[y^{-1} \log |u(iy)F(iy)|_{\mathscr{B}(\mathscr{C})} - y^{-1} \log |u(iy)| \right] \\
&\leq -\lim_{y \to \infty} y^{-1} \log |u(iy)| \\
&= m,
\end{aligned}
$$

where m is a finite real constant. Thus $\tau_c \leq m < \infty$ for every $c \in \mathscr{C}$ and $\tau < \infty$. If $\tau = -\infty$, then $\tau_c = -\infty$ for every $c \in \mathscr{C}$. Hence $F_c \equiv 0$ for every $c \in \mathscr{C}$ and $F \equiv 0$. ∎

6.10   Entire Functions of Exponential Type

If F is a $\mathscr{B}(\mathscr{C})$-valued entire function, define F_c for any $c \in \mathscr{C}$ by

$$F_c(z) = \langle F(z)c, c \rangle_{\mathscr{C}}, \qquad z \in \mathbf{C}.$$

DEFINITION.   *A* $\mathscr{B}(\mathscr{C})$-*valued entire function F is of* exponential type *if there is a real constant m such that for each* $c \in \mathscr{C}$,

$$|F_c(z)| \leq M_c e^{m|z|}, \qquad z \in \mathbf{C},$$

for some $M_c > 0$. *In this case, the* exact type τ_F *of F is the infimum of all such m. Equivalently,*

$$\tau_F = \sup_{c \in \mathscr{C}} \left(\limsup_{|z| \to \infty} \frac{\log |\langle F(z)c, c \rangle_{\mathscr{C}}|}{|z|} \right).$$

We say that F is of exponential type τ *if F is of exponential type and* $\tau_F \leq \tau$.

It is easy to see that F is of exponential type τ if and only if F_c is of exponential type τ for each $c \in \mathscr{C}$. The exact type τ_F is the supremum of the exact types of all functions $F_c, c \in \mathscr{C}$. As in the scalar case, either $F \equiv 0$ and $\tau_F = -\infty$, or $F \not\equiv 0$ and $\tau_F \geq 0$.

6.11   Operator Extension of Kreĭn's Theorem

If F is a $\mathscr{B}(\mathscr{C})$-valued entire function, let $\tilde{F}$ be the reflection of F with respect to the real line: $\tilde{F}(z) = F(\bar{z})^*$, $z \in \mathbf{C}$. The following result generalizes Kreĭn's theorem in Section 5 of the Appendix.

THEOREM.   *If F is a $\mathscr{B}(\mathscr{C})$-valued entire function, the following are equivalent:*

(i) *F is of exponential type and*

$$\int_{-\infty}^{\infty} \frac{\log^+ |F(x)|_{\mathscr{B}(\mathscr{C})}}{1 + x^2} dx < \infty; \tag{6-16}$$

(ii) *the restrictions of F and $\tilde{F}$ to Π are of bounded type, that is, they belong to $N_{\mathscr{B}(\mathscr{C})}(\Pi)$.*

Let F satisfy these conditions, and let τ_+, τ_- be the mean types of the restrictions of $F, \tilde{F}$ to Π, respectively. Then

$$\tau_+ + \tau_- \geq 0 \tag{6-17}$$

and

$$|\tau_\pm| \leq \max(\tau_+, \tau_-) = \tau_F, \tag{6-18}$$

where τ_F is the exact type of F.

For a $\mathscr{B}(\mathscr{C})$-valued function F on Π or $\mathbf{C}$, define F_c for each $c \in \mathscr{C}$ as in Sections 6.9 and 6.10.

LEMMA.   *A $\mathscr{B}(\mathscr{C})$-valued holomorphic function F on Π belongs to $N_{\mathscr{B}(\mathscr{C})}^+(\Pi)$ if and only if:*

(i) *$F_c \in N^+(\Pi)$ for each $c \in \mathscr{C}$, and*
(ii) *the limit $F_0(x) = \lim_{y \downarrow 0} F(x + iy)$ exists in the weak operator topology a.e. on R, and*

$$\int_{-\infty}^{\infty} \frac{\log^+ |F_0(z)|_{\mathscr{B}(\mathscr{C})}}{1 + x^2} dx < \infty. \tag{6-19}$$

Proof of lemma.   Necessity follows from Section 4.3, Theorem C, and Section 4.6, Theorem B.

Conversely, assume that (i) and (ii) hold. By (6-19) there is a scalar valued outer function v_0 on R such that $1/|v_0| = \max(1, |F_0|_{\mathscr{B}(\mathscr{C})})$ a.e. on R. Then $|v_0| \leq 1$ and $|v_0 F_0|_{\mathscr{B}(\mathscr{C})} \leq 1$ a.e. on R. Multiplying $v_0(t)$ by $1/(t + i)$ if necessary, we can assume that

$$\int_{-\infty}^{\infty} |v_0(t) F_0(t)|_{\mathscr{B}(\mathscr{C})}^2 dt < \infty.$$

Let v be the outer function on Π whose boundary function is v_0. By (i), $vF_c \in N^+(\Pi)$ for each $c \in \mathscr{C}$. Since vF_c has a square summable boundary function, it belongs to $H^2(\Pi)$ and

$$\frac{1}{2\pi i} \int_{-\infty}^{\infty} \frac{\langle v_0(t)F_0(t)c,c\rangle_{\mathscr{C}}}{t - \bar{z}} dt = 0, \qquad z \in \Pi.$$

By the arbitrariness of c,

$$\frac{1}{2\pi i} \int_{-\infty}^{\infty} \frac{v_0(t)F_0(t)}{t - \bar{z}} dt = 0, \qquad z \in \Pi.$$

By Section 4.8, Theorem B, $v_0F_0 \in H^2_{\mathscr{B}(\mathscr{C})}(R) \subseteq N^+_{\mathscr{B}(\mathscr{C})}(R)$. It follows that v_0F_0 is the boundary function of some $G \in N^+_{\mathscr{B}(\mathscr{C})}(\Pi)$. For each $c \in \mathscr{C}$, the scalar valued functions vF_c and G_c belong to $N^+(\Pi)$ and have the same boundary function. Hence $vF_c \equiv G_c$ on Π. Therefore $vF \equiv G$. Since $G \in N^+_{\mathscr{B}(\mathscr{C})}(\Pi)$ and v is outer, $F \in N^+_{\mathscr{B}(\mathscr{C})}(\Pi)$. ∎

Proof of theorem.   Let F satisfy (i). Let τ_F be the exact type of F, and choose $m > \tau_F$. For each $c \in \mathscr{C}$ there is a constant $M_c > 0$ such that $|F_c(z)| \leq M_c e^{m|z|}$, $z \in C$. Claim: the restrictions of $e^{imz}F_c$ and $e^{imz}\tilde{F}_c$ to Π belong to $N^+(\Pi)$. This is trivial if $F_c \equiv 0$, so suppose $F_c \not\equiv 0$. By the scalar version of Kreĭn's theorem (Appendix, Section 5), the restrictions of F_c and $\tilde{F}_c$ to Π are of bounded type, and the mean types of these restrictions do not exceed m. Since F_c and $\tilde{F}_c$ are entire, the restrictions of F_c and $\tilde{F}_c$ to Π have no singular inner functions in their canonical factorizations (Appendix, Section 6). Thus

$$F_c = e^{-ipz}B_1 g \qquad \text{and} \qquad \tilde{F}_c = e^{-iqz}B_2 g \qquad (6\text{-}20)$$

on Π, where $p \leq m$, $q \leq m$, B_1 and B_2 are Blaschke products, and g is outer (the outer factor may be chosen the same in each case since F_c and $\tilde{F}_c$ have the same modulus on R). Therefore the restrictions of $e^{imz}F_c$ and $e^{imz}\tilde{F}_c$ to Π belong to $N^+(\Pi)$. This proves the claim.

In view of the claim just proved and (6-16), the lemma implies that the restrictions of $e^{imz}F$ and $e^{imz}\tilde{F}$ to Π belong to $N^+_{\mathscr{B}(\mathscr{C})}(\Pi)$. Hence the restrictions of F and $\tilde{F}$ to Π are of bounded type, that is, (ii) holds.

Conversely, let (ii) hold. By Section 4.6, Theorem B, F satisfies (6-16). Let τ_+, τ_- be the mean types of the restrictions of $F, \tilde{F}$ to Π. For any $c \in \mathscr{C}$, let τ_{c+}, τ_{c-} be the mean types of the restrictions of $F_c, \tilde{F}_c$ to Π. By the scalar version of Kreĭn's theorem (Appendix, Section 5), F_c is of exponential type and exact type equal to $\max(\tau_{c+}, \tau_{c-})$. Since

$$\tau_{\pm} = \sup_{c \in \mathscr{C}} \tau_{c\pm}, \qquad (6\text{-}21)$$

$\max(\tau_{c+}, \tau_{c-}) \leq \max(\tau_+, \tau_-)$ ($\tau_{\pm}$ are finite by 6.9). Hence F is of exponential type and exact type $\tau_F \leq \max(\tau_+, \tau_-)$. In particular, (i) holds.

If $\tau_F < \max(\tau_+, \tau_-)$, then by the first part of the proof, $\tau_{c\pm} \leq \tau_F$ for all $c \in \mathscr{C}$ and so $\tau_\pm \leq \tau_F < \max(\tau_+, \tau_-)$, a contradiction. Hence $\tau_F = \max(\tau_+, \tau_-)$. Since $\tau_{c+} + \tau_{c-} \geq 0$ for all $c \in \mathscr{C}$, by (6-21), $\tau_+ + \tau_- \geq 0$. By an elementary argument, this implies $|\tau_\pm| \leq \max(\tau_+, \tau_-)$, and the proof is complete.   ∎

6.12   Kreĭn Classes

Let $\tau_1 \geq 0$ and $\tau_2 \geq 0$. A $\mathscr{B}(\mathscr{C})$-valued entire function F is of *class* $\mathscr{K}(\tau_1, \tau_2)$ if the restriction of F to Π is of bounded type and mean type $\leq \tau_1$, and the restriction of $\tilde{F}$ to Π is of bounded type and mean type $\leq \tau_2$. The classes $\mathscr{K}(\tau_1, \tau_2)$ are called *Kreĭn classes*.

THEOREM.   *Let* $u(x) = e^{i\tau_1 x}$, $v(x) = e^{i\tau_2 x}$ *for all real* x. *Let* F_0 *be a weakly measurable* $\mathscr{B}(\mathscr{C})$-*valued function on* R *such that* $|F_0|_{\mathscr{B}(\mathscr{C})}$ *is integrable over every bounded interval. Then* F_0 *is of class* $\mathscr{M}(u,v)$ *if and only if* F_0 *is equal a.e. to the restriction to* R *of a* $\mathscr{B}(\mathscr{C})$-*valued entire function* F *of class* $\mathscr{K}(\tau_1, \tau_2)$.

For an analogous result in the circle case, see the example in Section 6.3.

Proof.   Let F_0 be of class $\mathscr{M}(u,v)$. By Section 6.4 and Section 6.3, Theorem B, F_0 is equal a.e. to the restriction to R of an entire function F such that the restrictions to Π of $e^{i\tau_1 z} F$ and $e^{i\tau_2 z} \tilde{F}$ belong to $N^+_{\mathscr{B}(\mathscr{C})}(\Pi)$. It follows that the restrictions to Π of F and $\tilde{F}$ are of bounded type and mean type at most τ_1 and τ_2, respectively; that is, F is of class $\mathscr{K}(\tau_1, \tau_2)$.

Conversely, let $F_0(x) = F(x)$ a.e. on R, where F is a $\mathscr{B}(\mathscr{C})$-valued entire function of class $\mathscr{K}(\tau_1, \tau_2)$. For c in $\mathscr{C}$, let F_c and $\tilde{F}_c$ be defined by

$$F_c(z) = \langle F(z)c, c \rangle_{\mathscr{C}} \qquad \text{and} \qquad \tilde{F}_c(z) = \langle \tilde{F}(z)c, c \rangle_{\mathscr{C}}$$

for $z \in C$. Then F_c is entire, and the restrictions of F_c and $\tilde{F}_c$ to Π have canonical factorizations of the form (6-20), where $p \leq \tau_1, q \leq \tau_2$. Hence the restrictions of $e^{i\tau_1 z} F_c$ and $e^{i\tau_2 z} \tilde{F}_c$ to Π belong to $N^+(\Pi)$. Since F satisfies (6-16) the lemma of 6.11 implies that the restrictions of $e^{i\tau_1 z} F$ and $e^{i\tau_2 z} \tilde{F}$ to Π belong to $N^+_{\mathscr{B}(\mathscr{C})}(\Pi)$. The boundary functions of these restrictions are uF_0 and vF_0^*. Hence $uF_0, vF_0^* \in N^+_{\mathscr{B}(\mathscr{C})}(R)$ and F_0 is of class $\mathscr{M}(u,v)$.   ∎

6.13   Operator Extension of Ahiezer's Theorem

We now apply the preceding results to generalize Ahiezer's theorem of 6.1.

THEOREM.   *Let* F *be a* $\mathscr{B}(\mathscr{C})$-*valued entire function of exponential type* $\tau, \tau \geq 0$, *such that* $F(x) \geq 0$ *for all real* x, *and*

$$\int\limits_{-\infty}^{\infty} \frac{\log^+ |F(x)|_{\mathscr{B}(\mathscr{C})}}{1 + x^2} dx < \infty.$$

Then $F = \tilde{G}G$ for some $\mathscr{B}(\mathscr{C})$-valued entire function G such that $e^{-i\tau z/2}G$ is of exponential type $\tau/2$ and the restriction of G to Π is an outer function.

Here $\tilde{G}(z) = G(\bar{z})^*, z \in \mathbf{C}$.

Proof. The function F is of class $\mathscr{K}(\tau,\tau)$. By Section 6.12, the restriction F_0 of F to R is of class $\mathscr{M}(v,v)$, where $v(x) = e^{i\tau x}$ for all real x. Hence by 6.5, $F_0 = G_0^*G_0$ a.e. on R, where G_0 is outer and of class $\mathscr{M}(1,v)$ on R. By Section 6.12, G_0 is the restriction to R of an entire function G of class $\mathscr{K}(0,\tau)$. Since $F_0 = G_0^*G_0$ a.e. on R, $F = \tilde{G}G$. Since G is of class $\mathscr{K}(0,\tau)$, $e^{-i\tau z/2}G$ is of class $\mathscr{K}(\tau/2,\tau/2)$ and hence of exponential type $\tau/2$ by Section 6.11. The restriction of G to Π is of bounded type and has boundary function G_0. Since G_0 is outer on R, the restriction of G to Π is an outer function on Π. ∎

The following result generalizes Szegö's theorem in 6.1.

6.14   Theorem. Operator Extension of Szegö's Theorem

Let F be a weakly measurable nonnegative $\mathscr{B}(\mathscr{C})$-valued function that has invertible values σ-a.e. on Γ or a.e. on R. In the circle case assume that

$$\log^+|F(e^{i\theta})|_{\mathscr{B}(\mathscr{C})} \quad and \quad \log^+|F(e^{i\theta})^{-1}|_{\mathscr{B}(\mathscr{C})} \in L^1(\sigma),$$

and in the case of the real line assume that

$$(1 + x^2)^{-1}\log^+|F(x)|_{\mathscr{B}(\mathscr{C})} \quad and \quad (1 + x^2)^{-1}\log^+|F(x)^{-1}|_{\mathscr{B}(\mathscr{C})} \in L^1(-\infty,\infty).$$

Then $F = G^*G$ σ-a.e. on Γ or a.e. on R for some $\mathscr{B}(\mathscr{C})$-valued outer function G on Γ or R, respectively.

Proof. We give the proof in the circle case. The other case follows by a change of variables.

We first reduce to the case in which F is bounded. Introduce $F_1 = F/f$, where $f = \max(1,|F|_{\mathscr{B}(\mathscr{C})})$ on Γ. Since $\log f \in L^1(\sigma)$, $f = |g|^2$ for some outer function g on Γ. If $F_1 = G_1^*G_1$ σ-a.e. on Γ for some $\mathscr{B}(\mathscr{C})$-valued outer function G_1, then $F = G^*G$ σ-a.e. on Γ, where $G = gG_1$ is outer. Since F_1 satisfies the hypotheses of the theorem and is bounded, we may assume without loss of generality that F is bounded.

We apply 3.8 to the Toeplitz operator $T_2 = T(F)$ induced on $H_{\mathscr{C}}^2(\Gamma)$ by F as in 6.2. Choose $T_1 = T(\phi I_{\mathscr{C}})$, where $\phi(e^{i\theta}) = 1/|F(e^{i\theta})^{-1}|_{\mathscr{B}(\mathscr{C})}$ σ-a.e. on Γ. By an elementary argument, $\phi I_{\mathscr{C}} \leq F$ σ-a.e. on Γ. Therefore $T_1 \leq T_2$ by 6.2, Theorem B(i). We check the hypotheses (i) and (ii) of 3.8.

(i) Our assumptions imply that $\log \phi \in L^1(\sigma)$, so $\phi = |\psi|^2$ for some $\psi \in H^\infty(\Gamma)$. If A_1 is multiplication by $\psi I_{\mathscr{C}}$ on $H_{\mathscr{C}}^2(\Gamma)$, then A_1 is analytic and $T_1 = A_1^*A_1$.

(ii) Let $\{f_n\}_1^\infty$ be a sequence in $H_{\mathscr{C}}^2(\Gamma)$ such that

$$\lim_{n,k \to \infty} \langle T_2(f_n - f_k),f_n - f_k\rangle_2 = 0 \tag{6-22}$$

and

$$\lim_{n \to \infty} \langle T_1 f_n,f_n\rangle_2 = 0. \tag{6-23}$$

By (6-22), $\{F^{1/2}f_n\}_1^\infty$ is a Cauchy sequence in $L_\mathscr{C}^2(\sigma)$, and by (6-23), $\{\psi f_n\}_1^\infty$ converges to zero in the metric of $L_\mathscr{C}^2(\sigma)$. Here $\phi = |\psi|^2$ as above. Then $|\psi f_{n_j}|_\mathscr{C} \to 0$ σ-a.e. on Γ for some subsequence. Hence $|f_{n_j}|_\mathscr{C} \to 0$ and $|F^{1/2}f_{n_j}|_\mathscr{C} \to 0$ σ-a.e. on Γ. Therefore $F^{1/2}f_n \to 0$ in the metric of $L_\mathscr{C}^2(\sigma)$, and so

$$\lim_{n \to \infty} \langle T_2 f_n, f_n \rangle_2 = 0.$$

We have shown that the hypotheses of 3.8 are satisfied.

By 3.8, $T_2 = A_2^* A_2$ for some analytic operator A_2 on $H_\mathscr{C}^2(\Gamma)$. By Section 5.12, Theorem C, A_2 is multiplication by G for some $G \in H_{\mathscr{B}(\mathscr{C})}^\infty(\Gamma)$. By 6.2, Theorem B(ii), $F = G^*G$ σ-a.e. on Γ. By 5.7 we may choose G to be outer, and this completes the proof. ∎

Examples and Addenda

1. Matrix valued functions may be viewed as a special case of operator valued functions.

THEOREM. Let $F = \begin{bmatrix} u & v \\ \bar{v} & w \end{bmatrix}$ be a measurable 2×2 matrix valued function on Γ such that $F \geq 0$ σ-a.e. and $\log^+(u + w) \in L^1(\sigma)$. Then F has the form $F = G^*G$ σ-a.e., where $G = \begin{bmatrix} a & b \\ c & d \end{bmatrix}$ is a 2×2 matrix valued outer function on Γ if and only if one of the following alternatives holds:
  (i) F has rank 0 σ-a.e., that is, $F = 0$ σ-a.e.;
  (ii) F has rank 1 σ-a.e. and either (a) $\log u \in L^1(\sigma)$ and $v/u \in N(\Gamma)$, or (b) $\log w \in L^1(\sigma)$ and $\bar{v}/w \in N(\Gamma)$;
  (iii) F has rank 2 σ-a.e. and $\log(\det F) \in L^1(\sigma)$.

This result is given in Masani and Wiener [1959]. See Matveev [1959] for a similar result for matrix valued functions of arbitrary finite order.

Proof. We write " $=$ " and " $\neq$ " for " $= \sigma$-a.e." and " $\neq \sigma$-a.e."

Sufficiency. This is trivial in case (i), and case (iii) follows from 6.14. Assume (ii) and (a). Then $u = |a|^2$, where a is outer, and $v/u = c/b$, where $c \in N^+(\Gamma)$, b is inner, and b, c have no common nonconstant inner factor. Since F has rank 1 σ-a.e., $uw - |v|^2 = \det F = 0$, and hence $w = |ac|^2$. It is not hard to see that the required factorization is obtained with $G = \begin{bmatrix} ab & ac \\ 0 & 0 \end{bmatrix}$. The case (b) of (ii) is handled similarly.

Necessity. Let $F = G^*G$ as in the theorem. Since a function in $N(\Gamma)$ cannot vanish on a set of positive measure without vanishing σ-a.e., we see without difficulty that F has constant rank 0, 1, or 2 σ-a.e. For the rank 0 case, we obviously have alternative (i). For the rank 2 case, $\log(\det F) = 2\log|\det G| \in L^1(\sigma)$ by Szegö's theorem, and (iii) holds. In the rank 1 case,

$$ad - bc = 0, \qquad u = |a|^2 + |c|^2,$$

$$v = \bar{a}b + \bar{c}d, \qquad w = |b|^2 + |d|^2,$$

and either $|a|^2 + |c|^2 \neq 0$ or $|b|^2 + |d|^2 \neq 0$. Assume $|a|^2 + |c|^2 \neq 0$. Then either $a \neq 0$ or $c \neq 0$, and hence either

$$\frac{v}{u} = \frac{\bar{a}b + \bar{c}d}{\bar{a}a + \bar{c}c} = \frac{\bar{a}b + \bar{c}bc/a}{\bar{a}a + \bar{c}c} = \frac{b}{a} \in N(\Gamma)$$

or

$$\frac{v}{u} = \frac{\bar{a}b + \bar{c}d}{\bar{a}a + \bar{c}c} = \frac{\bar{a}ad/c + \bar{c}d}{\bar{a}a + \bar{c}c} = \frac{d}{c} \in N(\Gamma).$$

Thus the alternative (a) of (ii) holds. The case $|b|^2 + |d|^2 \neq 0$ leads similarly to the alternative (b) of (ii). ∎

2. Let f be a scalar valued entire function of exponential type τ, $\tau \geq 0$, such that $|f(x)| \leq 1$ for all real x. Then $f = ab + cd$, where a,b,c,d are entire functions of exponential type $\frac{1}{2}\tau$ such that $\tilde{a}a + \tilde{c}c = \tilde{b}b + \tilde{d}d = 1$. If $\tilde{f}f$ is not identically 1, then the representation may be chosen such that in addition $\tilde{a}d - b\tilde{c} \neq 0$ on Π. (Rosenblum and Rovnyak [1971]. For any entire function g, $\tilde{g}(z) = \bar{g}(\bar{z})$ for every complex number z.)

Notes

Physical and mathematical applications have spurred much of the interest in factorization problems for operator valued functions. Some of these applications are indicated in the literature cited below.

6.3–6.5   Theorem 6.5 is due to Rosenblum and Rovnyak [1971]. A generalization of this theorem has been given by V. I. Macaev; see Arov [1979a], p. 114. The operator extension of Carleman's theorem in Section 6.4 includes refinements due to Netherton [1973]. Beurling [1972] obtains a different criterion for analytic continuation across a real interval.

Douglas and Helton [1973] prove a weaker version of Theorem 6.5 and give an application to the problem of Darlington synthesis in electrical network theory. See also Arov [1973], [1974], [1975] and Helton [1978].

In the scalar case, more general notions of a pseudomeromorphic function were introduced by Shapiro [1968] and studied from the point of view of generalized analytic continuation. These notions have proved to be useful in a variety of applications, such as Anderson and Clunie [1984], Arov [1978], Douglas, Shapiro, and Shields [1970], Kriete [1971], Kriete and Rosenblum [1972], Marshall [1976], and Tumarkin [1966].

6.6   Extensions of the Fejér-Riesz theorem to matrix valued functions were obtained by Rosenblatt [1958] and Helson [1964]. Gohberg [1964] proved an infinite dimensional operator Fejér-Riesz theorem under a complete continuity hypothesis. The general case of 6.6 is due to Rosenblum [1968].

6.7–6.8   The matrix case has been widely known in applications; see, for example, Rozanov [1958], [1960], Masani [1966], and Popov [1973]. The general case was given in Rosenblum and Rovnyak [1971].

6.9–6.13   The operator extension of Ahiezer's theorem is due to Rosenblum and Rovnyak [1969], [1970]. The present form was given by the authors [1971]; the operator extension of Kreĭn's theorem is new and simplifies the presentation.

6.14   A generalization of Szegö's theorem for matrix valued functions was announced by Zasuhin [1941]. Zasuhin's proof was flawed, but a valid proof was supplied by M. G. Kreĭn and presented in his lectures. Rozanov [1958] reproduces Kreĭn's proof. Independently, Wiener [1955] attempted to prove a similar result, but his argument was also flawed. Valid proofs were given by Wiener and

Masani [1957], Helson and Lowdenslager [1958], and Wiener and Akutowicz [1959]. Devinatz [1961] extended the matrix generalization of Szegö's theorem to operator valued functions; Lax [1961] independently announced the same result. Lowdenslager [1963] introduced the comparison method for factoring nonnegative operator valued functions. Lowdenslager's method is used by Douglas [1966], who corrected a technical error in the method, and Sz.-Nagy and Foiaş [1970], and it is the basis for our proof of 6.14.

Additional Literature

PREDICTION THEORY FOR STATIONARY STOCHASTIC PROCESSES. Kolmogorov [1941], Wiener [1949], Wiener and Masani [1957], [1958], Helson and Lowdenslager [1958], [1961], Rozanov [1958], [1963], [1977], and Masani [1960], [1966].

MATHEMATICAL SYSTEMS THEORY AND ENGINEERING PROBLEMS. Arov [1971], [1973], [1974], [1975], [1979a,b], Kailath [1974], Kailath et al. [1977], Levinson et al. [1966], and Popov [1973].

ALGORITHMS FOR CONSTRUCTING FACTORS. References are given in Kailath [1974], p. 168.

WIENER-HOPF FACTORIZATION, SIGNED FACTORIZATION, FACTORIZATION OF OPERATORS RELATIVE TO A CHAIN OF SUBSPACES. Ball and Helton [1982a,b], Bart, Gohberg, and Kaashoek [1979], [1983], Clancey and Gohberg [1981], Gohberg and Kreĭn [1967], Jakubovič [1970], Nikolaĭčuk and Spĭtkovs'kiĭ [1975], Ran [1984], and Rozanov [1977].

POLYNOMIALS WHICH ARE POSITIVE ON A SYSTEM OF INTERVALS. Kreĭn, Levin, Nudel'man [1984].

Appendix

Topics in Function Theory

1. Subharmonic Functions and the Class $sh^1(D)$

A function u on a region $\Omega \subseteq \mathbf{C}$ to $[-\infty,\infty)$ is *upper semicontinuous* if for each $a \in \Omega$, $\lim \sup_{z \to a} u(z) \leq u(a)$.

THEOREM A.   *If $u: \Omega \to [-\infty,\infty)$ is upper semicontinuous, then the following conditions are equivalent:*

(i) *For each $a \in \Omega$ there is a disk $D(a,r_a) \subseteq \Omega$ such that*

$$u(a) \leq \frac{1}{2\pi} \int\limits_0^{2\pi} u(a + re^{i\theta})\, d\theta$$

whenever $0 < r < r_a$.

(ii) *For each $a \in \Omega$ there is a disk $D(a,r_a) \subseteq \Omega$ such that if $0 < r < r_a$ and p is a polynomial satisfying $u \leq \operatorname{Re} p$ on $\partial D(a,r)$, then $u \leq \operatorname{Re} p$ on $D(a,r)$.*

(iii) *For every open set A with compact closure $\bar{A} \subseteq \Omega$ and every continuous function $h: \bar{A} \to (-\infty,\infty)$ whose restriction to A is harmonic, if $u \leq h$ on ∂A, then $u \leq h$ on A.*

In this case the properties expressed in (i) and (ii) hold for all $a \in \Omega$ and all $r > 0$ such that $\bar{D}(a,r) \subseteq \Omega$.

An upper semicontinuous function $u: \Omega \to [-\infty,\infty)$ that satisfies the conditions in Theorem A is called *subharmonic* on Ω. In this section we sketch some results from the general theory of subharmonic functions and some special results for the unit disk D. Where no proofs are given the results are readily obtained from standard sources, such as those cited at the end of the section.

THEOREM B.   *Let u be subharmonic on $D(a,R)$, $u \not\equiv -\infty$. If $0 < r_1 < r_2 < R$, then*

$$-\infty < \frac{1}{2\pi} \int\limits_0^{2\pi} u(a + r_1 e^{i\theta})\, d\theta \leq \frac{1}{2\pi} \int\limits_0^{2\pi} u(a + r_2 e^{i\theta})\, d\theta.$$

Moreover, whether $u(a)$ is finite or $-\infty$,

$$\lim_{r \downarrow 0} \frac{1}{2\pi} \int\limits_0^{2\pi} u(a + re^{i\theta})\, d\theta = u(a).$$

Let u be subharmonic on a region Ω. Then u may take the value $-\infty$, but u cannot take this value too frequently unless $u \equiv -\infty$. For example, if $u \not\equiv -\infty$, then $\{z: u(z) = -\infty\}$ has two dimensional Lebesgue measure zero, and

$$\int_0^{2\pi} |u(a + re^{i\theta})| \, d\theta < \infty$$

whenever $\bar{D}(a,r) \subseteq \Omega$.

Subharmonic functions satisfy the maximum principle: if u is subharmonic on a region Ω and $u(z_0) \geq u(z)$ for some z_0 in Ω and all z in Ω, then u is constant in Ω.

THEOREM C.   *Let u be subharmonic on a region Ω. Let ϕ be nondecreasing and convex on $(-\infty,\infty)$, and set $\phi(-\infty) = \lim_{x \to -\infty} \phi(x)$. Then $\phi \circ u$ is subharmonic on Ω.*

Let $u \not\equiv -\infty$ be subharmonic on a region Ω. A *harmonic majorant* for u is any harmonic function h on Ω such that $u \leq h$ on Ω. A *least harmonic majorant* for u is any harmonic majorant h for u such that $h \leq k$ for every harmonic majorant k for u. If a least harmonic majorant exists, it is unique.

THEOREM D.   *Let $u \not\equiv -\infty$ be subharmonic on the unit disk D. Then u has a harmonic majorant if and only if*

$$\sup_{0 < r < 1} \frac{1}{2\pi} \int_0^{2\pi} u(re^{i\theta}) \, d\theta < \infty.$$

In this case u has a least harmonic majorant given by

$$h(z) = \lim_{r \uparrow 1} \frac{1}{2\pi} \int_0^{2\pi} P(z/r, e^{i\theta}) u(re^{i\theta}) \, d\theta \tag{A-1}$$

for all $z \in D$.

By $h^1(D)$ we mean the class of all real valued harmonic functions h on D satisfying the equivalent conditions (Duren [1970]):

(i) $\displaystyle\sup_{0 < r < 1} \frac{1}{2\pi} \int_0^{2\pi} |h(re^{i\theta})| \, d\theta < \infty;$

(ii) $h = h_+ - h_-$, where h_+, h_- are nonnegative harmonic functions on D;

(iii) there is a real Borel measure μ on Γ such that

$$h(z) = \int_\Gamma P(z, e^{it}) \, d\mu(e^{it}), \qquad z \in D.$$

(iv) $|h| \leq k$ for some nonnegative harmonic function k on D.

The situation in Theorem D becomes particularly interesting when the least harmonic majorant h for u belongs to $h^1(D)$. The next result characterizes this case. See also the Szegö-Solomentsev theorem in Section 2.

THEOREM E.   Let $u \not\equiv -\infty$ be subharmonic on D. The following conditions are equivalent:

(i) u has a harmonic majorant which belongs to $h^1(D)$;
(ii) u has a nonnegative harmonic majorant;
(iii) for some r_0, $0 < r_0 < 1$,

$$\sup_{r_0 < r < 1} \frac{1}{2\pi} \int_0^{2\pi} |u(re^{it})|\, dt < \infty;$$

(iv) if $u^+ = \max(u,0)$, then

$$\sup_{0 < r < 1} \frac{1}{2\pi} \int_0^{2\pi} u^+(re^{it})\, dt < \infty.$$

If these conditions hold, then the least harmonic majorant h for u belongs to $h^1(D)$. It can be calculated from either $(A\text{-}1)$ or the formula

$$h(z) = \int_\Gamma P(z,e^{it})\, d\mu(e^{it}), \qquad z \in D,$$

where $d\mu$ is the weak limit of the measures $u(re^{it})\, d\sigma$, that is,

$$\int_\Gamma f\, d\mu = \lim_{r \uparrow 1} \int_\Gamma f(e^{it}) u(re^{it})\, d\sigma(e^{it}) \qquad \text{(A-2)}$$

for every $f \in C(\Gamma)$.

DEFINITION.   The class of subharmonic functions characterized in Theorem E is denoted $sh^1(D)$.

Proof of Theorem E.   (i) $\Leftrightarrow$ (ii) This is clear from the fact that $h^1(D)$ consists of all differences $h = h_+ - h_-$, where h_+, h_- are nonnegative harmonic functions on D.

(iii) $\Leftrightarrow$ (iv) Since $u^+ \leq |u|$, (iii) implies (iv). Since $u^+ = \frac{1}{2}(|u| + u)$, for any $r \in (0,1)$,

$$\frac{1}{2\pi} \int_0^{2\pi} |u(re^{it})|\, dt = \frac{1}{\pi} \int_0^{2\pi} u^+(re^{it})\, dt - \frac{1}{2\pi} \int_0^{2\pi} u(re^{it})\, dt.$$

By Theorem B the second integral on the right is a nondecreasing function of r in $(0,1)$, and so (iv) implies (iii).

(i) $\Rightarrow$ (iv) Given (i), then

$$u(z) \le \int_\Gamma P(z,e^{it})\,d\mu(e^{it}), \qquad z \in D,$$

for some real Borel measure μ on Γ. If $\mu = \mu_+ - \mu_-$, where μ_+ and μ_- are nonnegative measures, then

$$u^+(z) \le \int_\Gamma P(z,e^{it})\,d\mu_+(e^{it}), \qquad z \in D.$$

Hence for any $r \in (0,1)$,

$$\int_\Gamma u^+(re^{i\theta})\,d\sigma(e^{i\theta}) \le \int_\Gamma \int_\Gamma P(re^{i\theta},e^{it})\,d\sigma(e^{i\theta})\,d\mu_+(e^{it})$$

$$= \mu_+(\Gamma),$$

and (iv) follows.

(iii) $\Rightarrow$ (i) Suppose that (iii) holds. By Theorem D, u has a least harmonic majorant given by (A-1). By (iii) and elementary estimates we can write (A-1) in the form

$$h(z) = \lim_{r\uparrow 1} \int_\Gamma P(z,e^{it})u(re^{it})\,d\sigma(e^{it}). \tag{A-3}$$

We may think of the measures $\{u(re^{it})\,d\sigma\}_{r_0 < r < 1}$ as linear functionals on $C(\Gamma)$. By (iii) and Alaoglu's theorem, for any sequence $r_n \uparrow 1$ there is a subsequence $r_{n(1)}, r_{n(2)}, \ldots$ and a real Borel measure μ such that

$$\lim_{k\to\infty} \int_\Gamma f(e^{it})u(r_{n(k)}e^{it})\,d\sigma(e^{it}) = \int_\Gamma f\,d\mu \tag{A-4}$$

for each $f \in C(\Gamma)$. Choosing $f(e^{it}) = P(z,e^{it})$ for any $z \in D$, we obtain

$$h(z) = \lim_{k\to\infty} \int_\Gamma P(z,e^{it})u(r_{n(k)}e^{it})\,d\sigma(e^{it})$$

$$= \int_\Gamma P(z,e^{it})\,d\mu(e^{it})$$

by (A-3) and (A-4). It follows that μ is independent of the choice of sequence $r_1, r_2, \ldots$ and subsequence $r_{n(1)}, r_{n(2)}, \ldots$. This proves not only (i) but also the last assertion of the theorem. ∎

Notes. The notation $sh^1(D)$ for the class of subharmonic functions characterized in Theorem E is not standard in the literature, though the class itself is well known. Concerning the theory of subharmonic functions and applications in analysis see, for example, Hayman and Kennedy [1976], Heins [1962], Hörmander [1973], Littlewood [1947], Radó [1949], and Tsuji [1959].

2. Theorem of Szegö-Solomentsev

The notation and assumptions that we set down here will be fixed throughout this section.

Let u be a nonnegative function on D, $u \not\equiv 0$, such that $\log u$ is subharmonic on D and

$$\sup_{0 < r < 1} \frac{1}{2\pi} \int_0^{2\pi} \log^+ u(re^{it}) \, dt < \infty. \tag{A-5}$$

It is often the case in applications that u itself is subharmonic, but we do not assume this. By Theorem E of Section 1, $\log u$ has a least harmonic majorant given by

$$h(z) = \int_\Gamma P(z, e^{it}) \, d\mu(e^{it}), \qquad z \in D, \tag{A-6}$$

where μ is a real Borel measure on Γ such that

$$\int_\Gamma f \, d\mu = \lim_{r \uparrow 1} \int_\Gamma f(e^{it}) \log u(re^{it}) \, d\sigma(e^{it}) \tag{A-7}$$

for each $f \in C(\Gamma)$. The theorem of Szegö-Solomentsev relates the growth of u to the structure of h.

By Fatou's theorem (Hoffman [1962], p. 34),

$$h(e^{i\theta}) = \lim_{z \to e^{i\theta}} h(z) \tag{A-8}$$

exists nontangentially σ-a.e. on Γ, and $h(e^{i\theta})$ is the density of the absolutely continuous component of the Lebesgue decomposition of μ:

$$d\mu = h(e^{i\theta}) \, d\sigma + d\lambda. \tag{A-9}$$

Let

$$\lambda = \lambda_+ - \lambda_- \tag{A-10}$$

be the Jordan decomposition of the singular component in (A-9).

Finally, we set

$$u(e^{i\theta}) = \lim_{r \uparrow 1} u(re^{i\theta}) \tag{A-11}$$

σ-a.e. Notice that (A-11) is a radial limit, whereas (A-8) is a nontangential limit. The existence of a finite and positive limit (A-11) σ-a.e. is clear in our applications, and there is no harm in making this a hypothesis for what follows.

Actually the existence of a finite and positive limit (A-11) σ-a.e. in the general case follows by a theorem of Littlewood [1928]. It is known that the radial limit in (A-11) cannot always be replaced by a nontangential limit as in (A-8). See Tsuji [1959], pp. 172–175, and Tolsted [1950], [1957].

THEOREM (*Szegö* [1921], *Solomentsev* [1938]). (i) *The boundary functions* (A-8) *and* (A-11) *are related by*

$$\log u(e^{i\theta}) = h(e^{i\theta}) \quad \sigma\text{-a.e.} \tag{A-12}$$

In particular, $\log u(e^{i\theta}) \in L^1(\sigma)$.

(ii) *For all* $z \in D$,

$$u(z) \le \frac{\exp\left(\int_\Gamma P(z,e^{it})h(e^{it})\,d\sigma\right)\exp\left(-\int_\Gamma P(z,e^{it})\,d\lambda_-\right)}{\exp\left(-\int_\Gamma P(z,e^{it})\,d\lambda_+\right)}. \tag{A-13}$$

(iii) *If there is a function* $\phi(t) \ge 0$ *of* $t \ge 0$ *such that* $\phi(t)/t \to \infty$ *as* $t \to \infty$ *and for some open arc* γ *of* Γ *(possibly* $\gamma = \Gamma$*),*

$$\sup_{0 < r < 1} \int_\gamma \phi(\log^+ u(re^{it}))\,d\sigma < \infty, \tag{A-14}$$

then $\lambda_+|\gamma = 0$.

(iv) *If in* (iii) *we replace* (A-14) *by the stronger condition*

$$\sup_{\frac{1}{2} < r < 1} \int_\gamma \phi(|\log u(re^{it})|)\,d\sigma < \infty, \tag{A-15}$$

then $\lambda_+|\gamma = \lambda_-|\gamma = 0$.

Proof. (i) Clearly $\log u(e^{i\theta}) \le h(e^{i\theta})$ σ-a.e. since $\log u \le h$ on D. For any sequence $r_n \uparrow 1$,

$$\int_\Gamma [h(e^{i\theta}) - \log u(e^{i\theta})]\,d\sigma \le \lim_{n \to \infty} \int_\Gamma [h(r_n e^{i\theta}) - \log u(r_n e^{i\theta})]\,d\sigma$$

$$= \lim_{n \to \infty} \left[h(0) - \int_\Gamma \log u(r_n e^{i\theta})\,d\sigma \right] = 0$$

by Fatou's lemma and Theorem D of Section 1. Thus (A-12) follows.

(ii) The estimate (A-13) is an exponentiated form of the inequality $\log u \le h$.

(iv) Suppose that (A-15) holds. Let $\varepsilon > 0$ be given. By the theorem of de la Vallée Poussin and Nagumo in Section 3, there is a $\delta > 0$ such that

$$\int_\Delta |\log u(re^{it})|\,d\sigma < \varepsilon/2, \quad \tfrac{1}{2} < r < 1,$$

for every Borel set $\Delta \subseteq \gamma$ with $\sigma(\Delta) < \delta$. Let Δ be an open subset of γ with $\sigma(\Delta) < \delta$. Consider any $f \in C(\Gamma)$, $|f| \leq 1$, such that $f = 0$ on $\Gamma\setminus\Delta$. For any $r \in (\frac{1}{2}, 1)$,

$$\left|\int_\Delta f \, d\mu\right| \leq \left|\int_\Gamma f \, d\mu - \int_\Gamma f(e^{it}) \log u(re^{it}) \, d\sigma\right|$$

$$+ \left|\int_\Delta f(e^{it}) \log u(re^{it}) \, d\sigma\right|$$

$$< \left|\int_\Gamma f \, d\mu - \int_\Gamma f(e^{it}) \log u(re^{it}) \, d\sigma\right| + \frac{\varepsilon}{2}.$$

By Theorem E of Section 1 we can choose r such that the first term on the right does not exceed $\varepsilon/2$. Thus $|\int_\Delta f \, d\mu| < \varepsilon$. Since f is arbitrary, $|\mu(\Delta)| \leq \varepsilon$. Therefore $\mu|\gamma$ is absolutely continuous, so $\lambda|\gamma = 0$ and (iv) follows.

(iii) Let (A-14) hold. Write

$$\log u = \log^+ u - \log^- u, \qquad |\log u| = \log^+ u + \log^- u,$$

where $\log^\pm u = \max(\pm\log u, 0)$. By (A-5) and Theorem E of Section 1,

$$\sup_{r_0 < r < 1} \int_\Gamma \log^\pm u(re^{it}) \, d\sigma < \infty,$$

where $0 < r_0 < 1$. Applying Alaoglu's theorem as in the proof of Theorem E of Section 1, we obtain measures $v_+ \geq 0$ and $v_- \geq 0$ on Γ and a sequence $r_n \uparrow 1$ such that

$$\lim_{n \to \infty} \int_\Gamma f(e^{it}) \log^\pm u(r_n e^{it}) \, d\sigma = \int_\Gamma f \, dv_\pm$$

for all $f \in C(\Gamma)$. Hence by the last assertion in Theorem E of Section 1, $\mu = v_+ - v_-$. It also follows from Theorem E of Section 1 that $\log^+ u = \log \max(u, 1)$ belongs to $sh^1(D)$ and has the least harmonic majorant

$$h^+(z) = \int_\Gamma P(z, e^{it}) \, dv_+, \qquad z \in D.$$

Applying part (iv) of the theorem, which has already been proved, to $\log^+ u$, we see that $v_+|\gamma$ is absolutely continuous. Hence if N is a Borel subset of γ with $\sigma(N) = 0$, then $\mu(N) = -v_-(N) \leq 0$. Therefore $\lambda|\gamma \leq 0$ and $\lambda_+|\gamma = 0$. ∎

Notes. The paper of Szegö [1921] is one of the sources of modern function theory. The work of Solomentsev [1938] remained little known outside the Soviet Union until it was rediscovered by Gårding and Hörmander [1964].

Heins [1967], [1969], [1983] recognized its great power for generalizations of classical function theory to Riemann surfaces and operator valued functions. A similar method is used in Rudin [1969]. The result was applied to the study of operator valued functions in Rosenblum and Rovnyak [1971] (see Chapter 4).

3. Strongly Convex Functions and a Theorem of de la Vallée Poussin and Nagumo

A function ϕ on $(-\infty,\infty)$ is *strongly convex* if (i) ϕ is convex, (ii) ϕ is nondecreasing, (iii) $\phi \geq 0$, (iv) $\phi(t)/t \to \infty$ as $t \to \infty$, and (v) for some $c > 0$ there exist constants $M \geq 0$ and $a \in (-\infty,\infty)$ such that $\phi(t + c) \leq M\phi(t)$ for all $t \geq a$.

If (v) holds for one value of $c > 0$, then it holds for all $c > 0$. An equivalent form of (v) is: (v') for every $c \in (-\infty,\infty)$ there exist contants $M \geq 0$ and $K \geq 0$ such that $\phi(t + c) \leq M\phi(t) + K$ for all real t.

Each of the following functions is strongly convex:

$$\phi(t) = e^{pt}, \qquad t \text{ real}, \qquad 0 < p < \infty,$$

$$\phi(t) = \begin{cases} t^p, & t \geq 0, \quad 1 < p < \infty, \\ 0, & t < 0, \end{cases}$$

$$\phi(t) = \begin{cases} t \log t, & t \geq 1, \\ 0, & t < 1. \end{cases}$$

Let $(A,\mathscr{F},\mu)$ be a nonnegative measure space. A family of functions $\{f_j\}_{j \in J}$ in $L^1(\mu)$ is *uniformly integrable* if:

(i) $\displaystyle\sup_{j \in J} \int_A |f_j| \, d\mu < \infty;$

(ii) for every $\varepsilon > 0$ there is a $\delta > 0$ such that

$$\int_\Delta |f_j| \, d\mu < \varepsilon, \qquad j \in J,$$

for all $\Delta \in \mathscr{F}$ with $\mu(\Delta) < \delta$.

THEOREM (*de la Vallée Poussin* [1915], *Nagumo* [1929]). *Let* $\{f_j\}_{j \in J}$ *be any family in* $L^1(\mu)$. *If* $\mu(A) = \infty$, *we assume also that* $\int_A |f_j| \, d\mu \leq M$ *for all* $j \in J$ *and some* $M > 0$.

(i) *If there is a function* $\phi(t) \geq 0$ *of* $t \geq 0$ *such that* $\phi(t)/t \to \infty$ *as* $t \to \infty$ *and*

$$\sup_{j \in J} \int_A \phi(|f_j|) \, d\mu < \infty, \qquad (A\text{-}16)$$

then $\{f_j\}_{j \in J}$ *is uniformly integrable.*

(ii) *If $\{f_j\}_{j\in J}$ is uniformly integrable, then there exists a strongly convex function ϕ such that (A-16) holds.*

Notes. Rudin [1969] uses only the first four of our conditions that define strongly convex functions. The fifth condition is used in Theorem B of Section 4.2 and Theorem C of Section 4.6. Nevertheless, the main construction needed to prove the theorem is identical to that of Rudin [1969], p. 38. See also Graves [1956], p. 235, and McShane [1944], p. 176.

4. A Theorem of Flett and Kuran

It is useful to have a simple condition for the existence of a harmonic majorant for a subharmonic function on a half-plane. For our purposes it is sufficient to consider only nonnegative functions.

THEOREM (*T. M. Flett, Ü. Kuran* [1971]). *A nonnegative subharmonic function $v(z)$ on the upper half-plane Π has a harmonic majorant on Π if and only if*

$$\sup_{y>0} \int_{-\infty}^{\infty} \frac{v(x+iy)}{x^2+(y+1)^2} dx < \infty. \tag{A-17}$$

LEMMA. *Let $g(x)$ be a nonnegative and nondecreasing function on $[0,1)$. Let $p(x)$ be any nonnegative measurable function on $(0,1)$ such that*

$$0 < \int_0^a p(t)\,dt < \infty \tag{A-18}$$

for every $a \in (0,1)$ and

$$\int_0^1 p(t)\,dt = \infty. \tag{A-19}$$

Then

$$\lim_{x\uparrow 1} g(x) = \sup_{0<\lambda<1} \int_0^1 g(t)p(\lambda t)\,dt \Big/ \int_0^1 p(\lambda t)\,dt,$$

where the possibility that both sides are infinite is allowed.

Proof. Set $q(\lambda) = \int_0^1 g(t)p(\lambda t)\,dt/\int_0^1 p(\lambda t)\,dt$ for $0 < \lambda < 1$. If $\lim_{x\uparrow 1} g(x) = M < \infty$, then $q(\lambda) \leq M$ for $0 < \lambda < 1$, and so

$$\lim_{x\uparrow 1} g(x) \geq \sup_{0<\lambda<1} q(\lambda). \tag{A-20}$$

This inequality holds trivially if the left side is infinite.

Suppose $\sup_{0<\lambda<1} q(\lambda) = K < \infty$ and $0 < x < \lambda < 1$. Then

$$K \ge q(\lambda) \ge \int_x^1 g(t)p(\lambda t)\, dt \bigg/ \int_0^1 p(\lambda t)\, dt$$

$$\ge g(x) \int_x^1 p(\lambda t)\, dt \bigg/ \int_0^1 p(\lambda t)\, dt = g(x)\left[1 - \int_0^x p(\lambda t)\, dt \bigg/ \int_0^1 p(\lambda t)\, dt\right].$$

Letting $\lambda \uparrow 1$ and using (A-18) and (A-19), we obtain $K \ge g(x)$. It follows that equality holds in (A-20). Equality holds trivially in (A-20) if the right side is infinite, and this proves the lemma. ∎

Proof of theorem. Suppose first that $v(z)$ has a harmonic majorant on Π. By the Poisson representation of a positive harmonic function (see Section 6),

$$v(z) \le cy + \frac{y}{\pi} \int_{-\infty}^{\infty} \frac{d\mu(t)}{(t-x)^2 + y^2}, \qquad y > 0,$$

where $c \ge 0$ and μ is a measure satisfying

$$\int_{-\infty}^{\infty} (1 + t^2)^{-1}\, d\mu(t) < \infty.$$

Therefore

$$\int_{-\infty}^{\infty} \frac{v(x+iy)}{x^2 + (y+1)^2}\, dx$$

$$\le \pi c \frac{y}{\pi} \int_{-\infty}^{\infty} \frac{dx}{x^2 + (y+1)^2}$$

$$+ \frac{1}{y+1} \int_{-\infty}^{\infty} \frac{y}{\pi} \int_{-\infty}^{\infty} \frac{1}{(t-x)^2 + y^2} \frac{y+1}{x^2 + (y+1)^2}\, dx\, d\mu(t)$$

$$= \frac{\pi c y}{y+1} + \frac{2y+1}{y+1} \int_{-\infty}^{\infty} \frac{d\mu(t)}{t^2 + (2y+1)^2}$$

$$< \pi c + 2 \int_{-\infty}^{\infty} \frac{d\mu(t)}{1 + t^2}$$

for all $y > 0$, and (A-17) follows.

Conversely assume that (A-17) holds. We use the mappings

$$\alpha: w \to i(1 + w)/(1 - w) \qquad \text{and} \qquad \beta: z \to (z - i)/(z + i).$$

Thus α maps D onto Π and $\beta = \alpha^{-1}$. Let $w = \beta(z)$, where $z = x + iy \in \Pi$ and $w = u + iv \in D$. Then area elements are transformed by the formula

$$du\, dv = |\beta'(z)|^2\, dx\, dy.$$

We must show that v has a harmonic majorant on Π, or, what is the same thing, that $v \circ \alpha$ has a harmonic majorant on D. The condition for this (see Section 1) is that

$$g(r) = r \int_\Gamma v(\alpha(re^{i\theta}))\, d\sigma(e^{i\theta})$$

remain bounded as $r \uparrow 1$. By the lemma it is sufficient to show that

$$\int_0^1 \frac{g(r)}{1 - \lambda^2 r^2}\, dr \le \text{const.} \int_0^1 \frac{dt}{1 - \lambda^2 t^2}$$

for all $\lambda \in (0,1)$ and some positive constant. Calculate as follows:

$$\int_0^1 \frac{g(r)}{1 - \lambda^2 r^2}\, dr = \frac{1}{2\pi} \int_0^1 \int_0^{2\pi} \frac{v(\alpha(re^{i\theta}))}{1 - \lambda^2 r^2}\, d\theta\, r\, dr$$

$$= \frac{1}{2\pi} \iint_D \frac{v(\alpha(w))}{1 - \lambda^2 |w|^2}\, du\, dv$$

$$= \frac{1}{2\pi} \iint_\Pi \frac{v(z)}{1 - \lambda^2 |\beta(z)|^2} |\beta'(z)|^2\, dx\, dy$$

$$= \frac{2}{\pi} \iint_\Pi \frac{v(z)}{|z + i|^2} \frac{dx\, dy}{|z + i|^2 - \lambda^2 |z - i|^2}$$

$$= \frac{2}{\pi} \iint_\Pi \frac{v(x + iy)}{x^2 + (y + 1)^2} \frac{dx\, dy}{(1 - \lambda^2)x^2 + (y + 1)^2 - \lambda^2(y - 1)^2}$$

$$\le \frac{2}{\pi} \int_0^\infty \left(\int_{-\infty}^\infty \frac{v(x + iy)}{x^2 + (y + 1)^2}\, dx \right) \frac{dy}{(y + 1)^2 - \lambda^2(y - 1)^2}.$$

By (A-17),

$$\int_0^1 \frac{g(r)}{1 - \lambda^2 r^2}\, dr \le \text{const.} \int_0^\infty \frac{dy}{(y+1)^2 - \lambda^2(y-1)^2}$$

$$= \text{const.} \int_{-1}^1 \frac{dt}{1 - \lambda^2 t^2} = \text{const.} \int_0^1 \frac{dt}{1 - \lambda^2 t^2}.$$

The change of variables is made with the subsitution $t = (y-1)/(y+1)$. The theorem follows. ∎

Notes. Kuran [1971] proves a more general result for nonnegative subharmonic functions on an $(n+1)$-dimensional half-space. According to Kuran, T. M. Flett proved the necessity part and conjectured the general result in private correspondence. In the special case of subharmonic functions of the form $\log^+ |F(z)|$, where $F(z)$ is holomorphic on Π, the theorem is due to Wishard [1942]; Franck [1952] gives a different proof in this case. A weaker theorem for nonnegative subharmonic functions is given by Krylov [1939].

5. A Theorem of M. G. Kreĭn

An entire function $F(z)$ is of *exponential type* if

$$|F(z)| \le M e^{m|z|}, \qquad z \in \mathbf{C},$$

for some real constant m and positive constant M. The *exact type* of $F(z)$ is then defined by

$$\tau_F = \limsup_{|z| \to \infty} |z|^{-1} \log|F(z)|.$$

THEOREM (*Kreĭn* [1947]). *If $F(z)$ is an entire function, then the following are equivalent:*

(i) *$F(z)$ is of exponential type and*

$$\int_{-\infty}^\infty \frac{\log^+ |F(t)|}{1 + t^2}\, dt < \infty; \qquad (A-21)$$

(ii) *the restrictions of $F(z)$ and $\tilde{F}(z) = \bar{F}(\bar{z})$ to the upper half-plane Π are of bounded type; that is, they belong to the class $N(\Pi)$ defined in Section 4.2.*

Suppose that $F(z)$ satisfies these conditions. Let τ_F be the exact type of $F(z)$, and set

$$\tau_\pm = \limsup_{y \to \infty} y^{-1} \log|F(\pm iy)|.$$

Then $\tau_+ + \tau_- \geq 0$ and

$$|\tau_\pm| \leq max(\tau_+, \tau_-) = \tau_F.$$

Proof. We will only show that (ii) implies (i). The rest may be deduced from standard results in Boas [1954] and Levin [1980].

Suppose that $F(z)$ satisfies (ii). Let z be any complex number, and let $R > |z|$. Estimate as follows:

$$\log|F(z)| \leq \frac{1}{2\pi} \int_0^{2\pi} \frac{R^2 - |z|^2}{|Re^{it} - z|^2} \log^+|F(Re^{it})| \, dt$$

$$\leq \frac{R^2 - |z|^2}{(R - |z|)^2} \frac{1}{2\pi} \int_0^{2\pi} \log^+|F(Re^{it})| \, dt$$

$$\leq \frac{R + |z|}{R - |z|} \frac{1}{2\pi R^2} \int_R^{2R} \int_0^{2\pi} \log^+|F(re^{it})| \, dt \, r dr$$

$$= \frac{R + |z|}{R - |z|} \frac{1}{2\pi R^2} \iint_A \log^+|F(w)| \, du \, dv$$

$$\leq \frac{R + |z|}{R - |z|} \frac{(2R + 1)^2}{2\pi R^2} \iint_A \frac{\log^+|F(w)|}{u^2 + (|v| + 1)^2} \, du \, dv.$$

Here $A = \{w : R < |w| < 2R\}$ and $w = u + iv$. The third inequality follows from the fact that

$$r \to \frac{r}{2\pi} \int_0^{2\pi} \log^+|F(re^{it})| \, dt$$

is a nondecreasing function of $r > 0$. By (ii), $\log^+|F(z)|$ and $\log^+|\tilde{F}(z)|$ have harmonic majorants on Π. Therefore by the theorem of Flett and Kuran in Section 4,

$$\iint_A \frac{\log^+|F(w)|}{u^2 + (|v| + 1)^2} \, du \, dv \leq \int_{-2R}^{2R} \int_{-\infty}^{\infty} \frac{\log^+|F(w)|}{u^2 + (|v| + 1)^2} \, du \, dv \leq CR,$$

where C is a constant. Hence

$$\log|F(z)| \leq \text{const.} \frac{R + |z|}{R - |z|} R.$$

Now choose $R = 2|z|$. Then our estimate yields $\log|F(z)| \leq$ const.$|z|$, and so $F(z)$ is of exponential type. Since (A-21) holds for any function of bounded type, we obtain (i).

Concerning the remaining assertions in the theorem, see Boas [1954], pp. 92–93, 132, and Levin [1980], p. 251. The last assertion in the theorem follows from properties of the Phragmén-Lindelöf indicator function given in both Boas [1954] and Levin [1980]. ∎

Notes. Kreĭn's theorem is stated in Boas [1954], p. 132, with a hint for part of the proof. A different proof is indicated in a series of exercises in de Branges [1968], p. 38. Pitt [1983] has given a generalization by yet another method. Our proof that (ii) implies (i) follows Pitt [1983], except that where he relies on estimates of Levinson and McKean [1964], we use the theorem of Flett and Kuran. See also Levin [1980], Chapter V, and Macaev [1960], where sufficient conditions are given for an entire function to belong to the class characterized in Kreĭn's theorem.

Kreĭn's theorem has found applications in spectral theory, mathematical systems theory, and weighted polynomial approximation (Kreĭn [1951], [1952], Arov [1971], [1975], and Pitt [1976], [1983]). An application to the factorization problem for nonnegative operator valued functions was given in Rosenblum and Rovnyak [1970], [1971] (see Chapter 6).

6. Miscellaneous Theorems from Half-Plane Function Theory

The results that we list here are from the theory of holomorphic functions of bounded type on a half-plane. As usual, Π denotes the open upper half-plane.

THEOREM A (*Poisson Representation*). *Every nonnegative harmonic function $V(z)$ on Π has a representation*

$$V(z) = cy + \frac{y}{\pi} \int_{-\infty}^{\infty} \frac{d\mu(t)}{(t-x)^2 + y^2}, \tag{A-22}$$

where $c \geq 0$ and μ is a nonnegative Borel measure on $(-\infty, \infty)$ such that

$$\int_{-\infty}^{\infty} \frac{d\mu(t)}{t^2 + 1} < \infty. \tag{A-23}$$

THEOREM B (*Nevanlinna Representation*). *Every holomorphic function $F(z)$ on Π such that $\operatorname{Im} F(z) \geq 0$ on Π has the form*

$$F(z) = b + cz + \frac{1}{\pi} \int_{-\infty}^{\infty} \left(\frac{1}{t-z} - \frac{t}{1+t^2} \right) d\mu(t), \tag{A-24}$$

where $b = \bar{b}$, $c \geq 0$, and μ is a nonnegative Borel measure satisfying (A-23).

Both theorems have converses. Every function of the form (A-22) is non-negative and harmonic on Π, and every function of the form (A-24) is holomorphic and satisfies Im $F(z) \geq 0$ on Π.

THEOREM C (*Stieltjes Inversion Formula*).   *In the situation of Theorem A,*

$$\lim_{y \downarrow 0} \int_a^b V(x + iy)\, dx = \mu((a,b)) + \tfrac{1}{2}\mu(\{a\}) + \tfrac{1}{2}\mu(\{b\})$$

whenever $-\infty < a < b < \infty$.

THEOREM D (*Fatou's Theorem*).   *Let*

$$V(z) = \frac{y}{\pi} \int_{-\infty}^{\infty} \frac{d\mu(t)}{(t - x)^2 + y^2}, \qquad y > 0,$$

where μ is a nonnegative Borel measure on $(-\infty,\infty)$ satisfying (A-23). *If*

$$d\mu = F(x)\, dx + d\mu_s$$

is the Lebesgue decomposition of μ, then

$$\lim_{z \to x} V(z) = F(x)$$

nontangentially a.e. on $(-\infty,\infty)$.

The *canonical* (or *Nevanlinna*) *factorization* of functions of bounded type on Π is given in Krylov [1939]. Briefly, this asserts that each function $F \not\equiv 0$ in $N(\Pi)$ has the form

$$F(z) = e^{-i\tau z}B(z)G(z)S_+(z)/S_-(z), \qquad \text{(A-25)}$$

where τ is a real number, B is a Blaschke product, S_+ and S_- have the form

$$S_\pm(z) = \exp\left[\frac{1}{\pi i} \int_{-\infty}^{\infty} \left(\frac{1}{t - z} - \frac{t}{1 + t^2}\right) d\mu_\pm(t)\right]$$

where μ_+ and μ_- are nonnegative singular and mutually singular Borel measures on $(-\infty,\infty)$ such that

$$\int_{-\infty}^{\infty} \frac{d\mu_\pm(t)}{t^2 + 1} < \infty,$$

and G is an outer function. If F has an analytic continuation across some open interval I on $(-\infty,\infty)$, then $\mu_+|I = \mu_-|I = 0$.

DEFINITION.   *The number τ in* (A-25) *is called the* mean type *of F. The mean type of the function identically zero is taken to be* $-\infty$.

THEOREM E.   *The mean type τ of any function F in $N(\Pi)$ is given by*

$$\tau = \limsup_{y \to \infty} y^{-1} \log |F(iy)|. \tag{A-26}$$

If F has no zeros in Π, then the limit superior in (A-26) *is a limit.*

Notes.   Proofs of the theorems in this section and additional results on half-plane function theory may be found in Boas [1954], de Branges [1968], Duren [1970], Hoffman [1962], and Krylov [1939]. See especially de Branges [1968], where the notion of mean type is emphasized and other formulas for mean type are given. We mention also Paley and Wiener [1934], which is one of the classic sources for half-plane function theory.

Bibliography

The MR numbers indicate citations in Mathematical Reviews.

Adamjan, V. M., Arov, D. Z., and Kreĭn, M. G. (a) "Analytic properties of Schmidt pairs for a Hankel operator and the generalized Schur-Takagi problem," *Mat. Sb. (N.S.)* **86** (128), (1971), 34–75; *Math. USSR-Sb.* **15** (1971), 31–73, MR 45 # 7505; (b) "Infinite Hankel block matrices and related problems of extension," *Izv. Akad. Nauk Armjan. SSR Ser. Mat.* **6** (1971), 87–112, MR 45 # 7506.

Agler, J. "The Arveson extension theorem and coanalytic models," *Integral Equations and Operator Theory* **5** (1982), 608–631, MR 84g:47011.

Ahiezer, N. I. "On the theory of entire functions of finite degree," *Dokl. Akad. Nauk SSSR* **63** (1948), 475–478, MR 10,289; *Theory of Approximation*, Frederick Ungar, New York (1956), MR 20 # 1872; *The Classical Moment Problem*, "Moscow (1961);" Hafner, New York (1965), MR 32 # 1518.

Allison, A. C., and Young, N. J. "Numerical algorithms for the Nevanlinna-Pick problem," *Numer. Math.* **42** (1983), 125–145.

Anderson, J. M., and Clunie, J. "Characterizing sets in L^p-spaces," *Complex Variables Theory Appl.* **3** (1984), 33–44.

Arov, D. Z. "Darlington's method in the study of dissipative systems," *Dokl. Akad. Nauk SSSR* **201** (1971), 559–562; *Soviet Physics Dokl.* **16** (1971/72), 954–956, MR 55 # 1127; "Darlington realization of matrix-valued functions," *Izv. Akad. Nauk SSSR Ser. Mat.* **37** (1973), 1299–1331; *Math. USSR-Izv.* **7** (1973), 1295–1326, MR 50 # 10287; "On unitary couplings with losses (scattering theory with losses)," *Funktsional. Anal. i Prilozhen.* **8** (1974), 5–22; *Functional Anal. Appl.* **8** (1974), 280–294, MR 50 # 10864; "The realization of a canonical system with dissipative boundary conditions at one end of a segment in terms of the coefficient of dynamic flexibility," *Sibirsk. Mat. Zh.* **16** (1975), 440–463; *Siberian Math. J.* **16** (1975), 335–352, MR 57 # 12872; "An approximation characteristic of functions of the class BП," *Funktsional. Anal. i Prilozhen.* **12** (1978), 70–71; *Functional Anal. Appl.* **12** (1978), 133–134, MR 58 # 11420; (a) "Stable dissipative linear stationary dynamical systems of scattering," *J. Operator Theory* **2** (1979), 95–126, MR 81g:47007; (b) "Optimal and stable passive systems," *Dokl. Akad. Nauk SSSR* **247** (1979), 265–268; *Soviet Math. Dokl.* **20** (1979), 676–680, MR 80k:93036.

Ball, J. A. "Rota's theorem for general functional Hilbert spaces," *Proc. Amer. Math. Soc.* **64** (1977), 55–61, MR 57 # 1161; "Interpolation problems of Pick-Nevanlinna and Loewner types for meromorphic matrix functions," *Integral Equations and Operator Theory* **6** (1983), 804–840.

Ball, J. A., and Helton, J. W. (a) "Factorization results related to shifts in an indefinite metric," *Integral Equations and Operator Theory* **5** (1982), 632–658, MR 84m:47047; (b) "Lie groups over the field of rational functions, signed spectral factorization, signed interpolation, and amplifier design," *J. Operator Theory* **8** (1982), 19–64, MR 84j:94033; "A Beurling-Lax theorem for the Lie group $U(m,n)$ which contains most classical interpolation theory," *J. Operator Theory* **9** (1983), 107–142, MR 84m:47046.

Bart, H., Gohberg, I., and Kaashoek, M. A. *Minimal Factorization of Matrix and Operator Functions, Operator Theory: Advances and Applications*, Vol. 1, Birkhäuser Verlag, Basel (1979), MR 81a:47001; "Wiener-Hopf factorization of analytic operator functions and realization," Vrije Universiteit Amsterdam, Report No. 231 (1983).

Beals, R. *Topics in Operator Theory*, University of Chicago Press, Chicago (1971), MR 42 # 5065; *Advanced Mathematical Analysis*, Springer-Verlag, New York (1973).

Beatrous, F., Jr. "H^∞ interpolation from a subset of the boundary," *Pacific J. Math.* **106** (1983), 23–31.

Bendat, J. S., and Sherman, S. "Monotone and convex operator functions," *Trans. Amer. Math. Soc.* **79** (1955), 58–71, MR 18,588.

Beurling, A. "On two problems concerning linear transformations in Hilbert space," *Acta Math.* **81** (1949), 239–255, MR 10,381; "Analytic continuation across a linear boundary," *Acta Math.* **128** (1972), 153–182, MR 52 # 14368.

Boas, R. P., Jr. *Entire Functions*, Academic Press, New York (1954), MR 16,914.

de Branges, L. "Some Hilbert spaces of entire functions. III," *Trans. Amer. Math. Soc.* **100** (1961), 73–115, MR 24 # A3289c; "Some Hilbert spaces of analytic functions. II, III," *J. Math. Anal. Appl.* **11** (1965), 44–72; *ibid.* **12** (1965), 149–186, MR 35 # 778,779; *Hilbert Spaces of Entire Functions*, Prentice-Hall, Englewood Cliffs, N. J. (1968), MR 37 # 4590.

de Branges, L., and Rovnyak, J. "The existence of invariant subspaces," *Bull. Amer. Math. Soc.* **70** (1964), 718–721; correction, *ibid.* **71** (1965), 396, MR 28 # 4329 and 29 # 6279; (a) *Square Summable Power Series*, Holt, Rinehart and Winston, New York (1966), MR 35 # 5909; (b) "Appendix on square summable power series," in *Perturbation Theory and Its Applications in Quantum Mechanics*, Calvin H. Wilcox (ed.), 347–392, Wiley, New York (1966), MR 39 # 6109.

Brodskiĭ, M. S. *Triangular and Jordan Representations of Linear Operators*, Moscow (1969); *Transl. Math. Monographs*, Vol. 32, Amer. Math. Soc., Providence, R.I. (1971), MR 48 # 904.

Brown, A., and Halmos, P. R. "Algebraic properties of Toeplitz operators," *J. Reine Angew. Math.* **213** (1963), 89–102, MR 28 # 3350.

Brown, A., Halmos, P. R., and Shields, A. L. "Cesàro operators," *Acta Sci. Math. (Szeged)* **26** (1965), 125–137, MR 32 # 4539.

Burckel, R. B. *An Introduction to Classical Complex Analysis*, Birkhäuser Verlag, Basel (1979), MR 81d:30001.

Butzer, P. L., and Nessel, R. J. *Fourier Analysis and Approximation*, Vol. I, Birkhäuser Verlag, Basel (1971), MR 58 # 23312.

Calderón, A., Spitzer, F., and Widom, H. "Inversion of Toeplitz matrices," *Illinois J. Math.* **3** (1959), 490–498, MR 22 # 12386.

Carathéodory, C. "Über den Variabilitätsbereich der Koeffizienten von Potenzreihen, die gegebene Werte nicht annehmen," *Math. Ann.* **64** (1907), 95–115; *Gesammelte Math. Schriften*, Vol. III, 54–77; "Über den Variabilitätsbereich der Fourierschen Konstanten von positiven harmonischen Funktionen," *Rend. Circ. Mat. Palermo* **32** (1911), 193–217; *Gesammelte Math. Schriften*, Vol. III, 78–110.

Carathéodory, C., and Fejér, L. "Über den Zusammenhang der Extremen von harmonischen Funktionen mit ihren Koeffizienten und über den Picard-Landauschen Satz," *Rend. Circ. Mat. Palermo* **32** (1911), 218–239; *Gesammelte Math. Schriften*, Vol. III, 111–138.

Carleman, T. *L'Intégral de Fourier et Questions qui s'y Rattachent*, Almqvist and Wiksell, Uppsala (1944), MR 7,248.

Chandler, J. D., Jr. "Extensions of monotone operator functions," *Proc. Amer. Math. Soc.* **54** (1976), 221–224, MR 52 # 15074.

Clancey, K., and Gohberg, I. *Factorization of Matrix Functions and Singular Integral Operators, Operator Theory: Advances and Applications,* Vol. 3, Birkhäuser Verlag, Basel (1981), MR 84a:47016.

Clark, D. N. "On the point spectrum of a Toeplitz operator," *Trans. Amer. Math. Soc.* **126** (1967), 251–266, MR 34 # 1845; "Hankel forms, Toeplitz forms and meromorphic functions," *Trans. Amer. Math. Soc.* **134** (1968), 109–116, MR 37 # 6668; "On a similarity theory for rational Toeplitz operators," *J. Reine Angew. Math.* **320** (1980), 6–31, MR 82h:47025.

Daleckiĭ, Yu. L. "Integration and differentiation of functions of hermitian operators depending on a parameter," *Uspehi Mat. Nauk (N.S.)* **12** (1957), 182–186; *Amer. Math. Soc. Transl.* (2) **16** (1960), 396–400, MR 19, 155.

Dautov, Sh. A., and Khudaĭberganov, G. "The Carathéodory-Fejér problem in multidimensional complex analysis," *Sibirsk. Mat. Zh.* **23** (1982), 58–64; *Siberian Math. J.* **23** (1982), 183–188, MR 84g:32040.

Davis, C. "A Schwarz inequality for convex operator functions," *Proc. Amer. Math. Soc.* **8** (1957), 42–44, MR 18,812; "Notions generalizing convexity for functions defined on spaces of matrices," *Proc. Sympos. Pure Math.,* Vol. 7, 187–201, Amer. Math. Soc., Providence, R. I. (1963), MR 27 # 5771.

Delsarte, Ph., Genin, Y., and Kamp, Y. (a) "Schur parametrization of positive definite block-Toeplitz systems," *SIAM J. Appl. Math.* **36** (1979), 34–46, MR 80d:15013; (b) "The Nevanlinna-Pick problem for matrix-valued functions," *SIAM J. Appl. Math.* **36** (1979), 47–61, MR 80h:30035; "On the role of the Nevanlinna-Pick problem in circuit and system theory," *Internat. J. Circuit Theory Appl.* **9** (1981), 177–187, MR 82d:94052.

Devinatz, A. "The factorization of operator valued functions," *Ann. of Math.* (2) **73** (1961), 458–495, MR 23 # A3997; "On Wiener-Hopf operators," in *Functional Analysis,* B. R. Gelbaum (ed.), Proc. Conf., University of California, Irvine (1966), 81–118, Thompson Book Co., Washington (1967), MR 36 # 6873.

Dixmier, J. *Les Algèbres d'Opérateurs dans l'Espace Hilbertien (Algèbres de von Neumann),* 2nd ed., Gauthier-Villars, Paris (1969), MR 50 # 5482.

Donoghue, W. F., Jr. *Monotone Matrix Functions and Analytic Continuation,* Springer-Verlag, New York (1974), MR 58 # 6279; "Monotone operator functions on arbitrary sets," *Proc. Amer. Math. Soc.* **78** (1980), 93–96, MR 81g:47025; "Another extension of Loewner's theorem," preprint (1983); *J. Math. Anal. Appl.* (to appear).

Douglas, R. G. "On factoring positive operator functions," *J. Math. Mech.* **16** (1966), 119–126, MR 35 # 782; *Banach Algebra Techniques in Operator Theory,* Academic Press, New York (1972), MR 50 # 14355; *Banach Algebra Techniques in the Theory of Toeplitz Operators, Regional Conference Series in Mathematics,* no. 15, Amer. Math. Soc., Providence, R. I. (1973), MR 50 # 14336.

Douglas, R. G., and Helton, J. W. "Inner dilations of analytic matrix functions and Darlington synthesis," *Acta Sci. Math. (Szeged)* **34** (1973), 61–67, MR 48 # 900.

Douglas, R. G., Shapiro, H. S., and Shields, A. "Cyclic vectors and invariant subspaces for the backward shift operator," *Ann. Inst. Fourier (Grenoble)* **20** (1970), fasc. 1, 37–76, MR 42 # 5088.

Duren, P. L. *Theory of H^p Spaces,* Academic Press, New York (1970), MR 42 # 3552.

Durszt, E., and Sz.-Nagy, B. "Remark to a paper of A. E. Frazho: 'Models for noncommuting operators,'" *J. Funct. Anal.* **52** (1983), 146–147, MR 84h:47011.

Dym, H., and McKean, H. P., Jr. *Fourier Series and Integrals,* Academic Press, New York

(1972), MR 56 #945; *Gaussian Processes, Function Theory, and the Inverse Spectral Problem*, Academic Press, New York (1976), MR 56 #6829.

Erwe, F. "Koeffizientenproblem für beschränkte Systeme holomorpher Funktionen und Verallgemeinerung eines Satzes von Carathéodory-Fejér," *Math. Z.* **93** (1966), 210–215, MR 34 #6073.

Faulkner, G. D., and Huneycutt, J. E., Jr. "Orthogonal decomposition of isometries in a Banach space," *Proc. Amer. Math. Soc.* **69** (1978), 125–128, MR 57 #3892.

Fedčina, I. P. "A criterion for the solvability of the Nevanlinna-Pick tangent problem," *Mat. Issled.* **7** (1972), no. 4 (26), 213–227, 258, MR 50 #4960; "The tangential Nevanlinna-Pick problem with multiple points," *Akad. Nauk Armjan. SSR Dokl.* **61** (1975), 214–218, MR 53 #13576.

Fillmore, P. A. *Notes on Operator Theory*, Van Nostrand, New York (1970), MR 41 #2414.

FitzGerald, C. H. "Quadratic inequalities and analytic continuation," *J. Analyse Math.* **31** (1977), 19–47, MR 58 #6189; "Conditions for a function on an arc to have a bounded analytic extension," Dept. of Math. Sciences, The Johns Hopkins University, Technical Report 303, July (1978), 48 pp.; "Particular cases of analytic extension of boundary values," *Complex Variables Theory Appl.* **3** (1984), 113–123.

FitzGerald, C. H., and Horn, R. A. "On quadratic and bilinear forms in function theory," *Proc. London Math. Soc.* (3) **44** (1982), 554–576, MR 83g:30039.

Foiaş, C. "Sur certains théorèmes de J. von Neumann concernant les ensembles spectraux," *Acta Sci. Math.* (*Szeged*) **18** (1957), 15–20, MR 19,757; "A remark on the universal model of G. C. Rota for contractions," *Com. Acad. R. P. Romîne* **13** (1963), 349–352, MR 31 #605.

Franck, A. "Analytic functions of bounded type," *Amer. J. Math.* **74** (1952), 410–422, MR 13,927.

Frazho, A. E. "Models for noncommuting operators," *J. Funct. Anal.* **48** (1982), 1–11, MR 84h:47010.

Gantmacher, F. R. *Theory of Matrices*, Vols. 1 and 2, Chelsea, New York (1959), MR 21 #6372c.

Gårding, L., and Hörmander, L. "Strongly subharmonic functions," *Math. Scand.* **15** (1964), 93–96; *ibid.* **18** (1966), 183, MR 31 #3621.

Garnett, J. B. *Bounded Analytic Functions*, Academic Press, New York (1981), MR 83g:30037.

Gellar, R., and Page, L. "Inner-outer factorizations of operators," *J. Math. Anal. Appl.* **61** (1977), 151–158, MR 58 #23727.

Ginzburg, Ju. P. (a) "Multiplicative representations of operator-functions of bounded form," *Uspehi Mat. Nauk SSSR* **22** (1967) (133), 163–165, MR 34 #6511; (b) "Divisors and minorants of operator-functions of bounded form," *Mat. Issled.* **2** (1967), 47–74; *Amer. Math. Soc. Transl.* (2) **103** (1974), 153–169, MR 39 #4701.

Gohberg, I. "The factorization problem for operator functions," *Izv. Akad. Nauk SSSR Ser. Mat.* **28** (1964), 1055–1082; *Amer. Math. Soc. Transl.* (2) **49** (1966), 130–161, MR 30 #5182.

Gohberg, I., and Kreĭn, M. G. "Systems of integral equations on a half-line with kernels depending on the difference of arguments," *Uspehi Mat. Nauk* (*N.S.*) **13** (1958), 3–72; *Amer. Math. Soc. Transl.* (2) **14** (1960), 217–287, MR 21 #1506; *Theory and Applications of Volterra Operators in Hilbert Space*, Moscow (1967); *Transl. Math. Monographs*, Vol. 24, Amer. Math. Soc., Providence, R. I. (1970), MR 41 #9041.

Gohberg, I., et al. Articles in commemoration of the one hundredth anniversary of the birth of Otto Toeplitz, *Integral Equations and Operator Theory* **4** (1981), 275–302, MR 82f:01102,01103.

Graves, L. M. *Theory of Functions of Real Variables*, 2nd ed., McGraw-Hill, New York (1956), MR 17,717.

Grenander, U., and Szegö, G. *Toeplitz Forms and Their Applications*, University of California Press, Berkeley, Calif. (1958), MR 20 # 1349.

Halmos, P. R. *Introduction to Hilbert Space*, Chelsea, New York (1951), MR 13,563; *Finite-Dimensional Vector Spaces*, 2nd ed., Van Nostrand, New York (1958), MR 19,725; "Shifts on Hilbert spaces," *J. Reine Angew. Math.* **208** (1961), 102–112, MR 27 # 2868; *A Hilbert Space Problem Book*, 2nd ed., Springer-Verlag, New York (1982), MR 84e:47001.

Hamburger, H. L., and Grimshaw, M. E. *Linear Transformations*, Cambridge University Press, London (1956), MR 12,836.

Hamilton, D. H. "Quadratic inequalities and interpolation," preprint (1983); *J. Analyse Math.* (to appear).

Hansen, F., and Pedersen, G. K. "Jensen's inequality for operators and Löwner's theorem," *Math. Ann.* **258** (1981/82), 229–241, MR 83g:47020.

Hartman, P. "On unbounded Toeplitz matrices," *Amer. J. Math.* **85** (1963), 59–78, MR 27 # 580.

Hartman, P., and Wintner, A. "On the spectra of Toeplitz's matrices," *Amer. J. Math.* **72** (1950), 359–366, MR 12,187; "The spectra of Toeplitz's matrices," *Amer. J. Math.* **76** (1954), 867–882, MR 17,499.

Hasumi, M. *Hardy Classes on Infinitely Connected Riemann Surfaces, Lecture Notes in Mathematics*, Vol. 1027, Springer-Verlag, New York (1983).

Hayman, W. K., and Kennedy, P. B. *Subharmonic Functions*, Academic Press, New York (1976), MR 57 # 665.

Heins, M. *Selected Topics in the Classical Theory of Functions of a Complex Variable*, Holt, Rinehart and Winston, New York (1962), MR 29 # 217; "On the theorem of Szegö-Solomentsev," *Math. Scand.* **20** (1967), 281–289, MR 39 # 4382; *Hardy Classes on Riemann Surfaces, Lecture Notes in Mathematics*, Vol. 98, Springer-Verlag, New York (1969), MR 40 # 338; "Vector-valued harmonic functions," L. Fejér-F. Riesz Centennial Symposium (Budapest, 1980), *Colloq. Math. Soc. János Bolyai* **35** (Functions, Series, Operators) (1983), 621–632; "A bibliography on Pick-Nevanlinna interpolation and cognate questions," preprint (1984).

Hellinger, E. D. "Spectra of quadratic forms in infinitely many variables," Northwestern University Studies I (1941), 133–172, MR 3,210.

Helson, H. *Lectures on Invariant Subspaces*, Academic Press, New York (1964), MR 30 # 1409.

Helson, H., and Lowdenslager, D. "Prediction theory and Fourier series in several variables. I, II," *Acta Math.* **99** (1958), 165–202; *ibid.* **106** (1961), 175–213, MR 20 # 4155 and 31 # 562.

Helton, J. W. "Orbit structure of the Möbius transformation semi-group acting in H^∞ (broadband matching)," in *Topics in Functional Analysis (essays dedicated to M. G. Kreĭn on the occasion on his 70th birthday)*, 129–157, *Adv. in Math. Suppl. Stud.*, Vol. 3, Academic Press, New York (1978), MR 81e:46019.

Hilbert, D. *Grundzüge einer allgemeinen Theorie der linearen Integralgleichungen*, Leipzig (1912); Chelsea, New York (1953), MR 15,37.

Hille, E., and Phillips, R. S. *Functional Analysis and Semi-Groups*, Amer. Math. Soc. Coll. Publ., Vol. 31, Providence, R. I. (1957), MR 19,664.

Hoffman, K. *Banach Spaces of Analytic Functions*, Prentice-Hall, Englewood Cliffs, N. J. (1962), MR 24 # A2844.

Hörmander, L. *An Introduction to Complex Analysis in Several Variables*, 2nd ed., North Holland, New York (1973), MR 49 # 9246.

Hunt, R., Muckenhoupt, B., and Wheeden, R. "Weighted norm inequalities for the conjugate function and Hilbert transform," *Trans. Amer. Math. Soc.* **176** (1973), 227–251, MR 47#701.

Iohvidov, I. S. *Hankel and Toeplitz Matrices and Forms*, Birkhäuser Verlag, Basel (1982), MR 83k:15021.

Ismagilov, R. S. "The spectrum of Toeplitz matrices," *Dokl. Akad. Nauk SSSR* **149** (1963), 769–772; *Soviet Math. Dokl.* **4** (1963), 462–465, MR 26#4190.

Jakubovič, V. A. "Factorization of symmetric matrix polynomials," *Dokl. Akad. Nauk SSSR* **194** (1970), 532–535; *Soviet Math. Dokl.* **11** (1970), 1261–1264, MR 42#6012.

Kac, M. "On the distribution of values of sums of type $\Sigma f(2^k t)$," *Ann. of Math.* (2) **47** (1946), 33–49, MR 7,436.

Kailath, T. "A view of three decades of linear filtering theory," *IEEE Trans. Inform. Theory* **IT-20** (1974), 146–181, MR 57#5337.

Kailath, T., et al. *Linear Least-Squares Estimation, Benchmark Papers in Electrical Engineering and Computer Science*, Vol. 17, Dowden, Hutchinson and Ross, Stroudsburg, Pa. (1977).

Kato, T. *Perturbation Theory for Linear Operators*, 2nd ed., Springer-Verlag, New York (1976), MR 53#11389.

Kolmogorov, A. N. "Stationary sequences in Hilbert's space," Bull. Moscow State Univ. 2, no. 6 (1941), 1–40, MR 13,1138.

Koosis, P. *Introduction to H_p Spaces, London Math. Soc. Lecture Note Series*, no. 40, Cambridge University Press, London (1980), MR 81c:30062.

Korányi, A. "Note on the theory of monotone operator functions," *Acta Sci. Math. (Szeged)* **16** (1955), 241–245, MR 19,126; "On a theorem of Löwner and its connections with resolvents of selfadjoint transformations," *Acta Sci. Math. (Szeged)* **17** (1956), 63–70, MR 18,588; "On some classes of analytic functions of several variables," *Trans. Amer. Math. Soc.* **101** (1961), 520–554, MR 25#226; "Monotone functions on formally real Jordan algebras," preprint (1983); *Math. Ann.* (to appear).

Korányi, A., and Pukánszky, L. "Holomorphic functions with positive real part on polycylinders," *Trans. Amer. Math. Soc.* **108** (1963), 449–456, MR 28#2247.

Kovalishina, I. V. "Analytic theory of a class of interpolation problems," *Izv. Akad. Nauk SSSR Ser. Mat.* **47** (1983), 455–497, MR 84i:30043.

Kovalishina, I. V., and Potapov, V. P. *Integral Representation of Hermitian Positive Functions*, Viniti, Kharkov (1982); private English transl. by T. Ando, Hokkaido Univ., Sapporo, Japan (1982).

Kraynek, W. T. "Interpolation of sublinear operators on generalized Orlicz and Hardy-Orlicz spaces," *Studia Math.* **43** (1972), 93–123, MR 47#807.

Kreĭn, M. G. "A contribution to the theory of entire functions of exponential type," *Izv. Akad. Nauk SSSR* **11** (1947), 309–326, MR 9,179; "On the theory of entire matrix-functions of exponential type," *Ukrain. Mat. Zh.* **3** (1951), 164–173, MR 14,981; "On the indeterminate case of a Sturm-Liouville boundary value problem in the interval $(0,\infty)$," *Izv. Akad. Nauk SSSR Ser. Mat.* **16** (1952), 293–324, MR 14,558; "Integral equations on a half-line with kernel depending on the difference of the arguments," *Uspehi Mat. Nauk (N.S.)* **13** (1958), 3–120; *Amer. Math. Soc. Transl.* (2) **22** (1962), 163–288, MR 21#1507.

Kreĭn, M. G., and Langer, H. "Über einige Fortsetzungsprobleme, die eng mit der Theorie hermitscher Operatoren im Raume Π_κ zusammenhängen. I. Einige Funktionenklassen und ihre Darstellungen," *Math. Nachr.* **77** (1977), 187–236, MR 57#1173.

Kreĭn, M. G., Levin, B. Ja., and Nudel'man, A. A. "On a special representation of a polynomial which is positive upon a system of closed intervals," Akad. Nauk Ukrain. SSR, Fiz.-Tech. Inst. Low Temp. (1984), 48 pp.

Kreĭn, M. G., and Nudel'man, A. A. *The Markov Moment Problem and Extremal Problems*, Moscow (1973); *Transl. Math. Monographs*, Vol. 50, Amer. Math. Soc., Providence, R. I. (1977), MR 56 # 16284.

Kreĭn, M. G., and Rehtman, P. G. "On the problem of Nevanlinna-Pick," *Trudi Odes'kogo Derz. Univ. Mat.* **2** (1938), 63–68.

Kriete III, T. L. "A generalized Paley-Wiener theorem," *J. Math. Anal. Appl.* **36** (1971), 529–555, MR 44 # 5473.

Kriete III, T. L., and Rosenblum, M. "A Phragmén-Lindelöf theorem with applications to $\mathscr{M}(u,v)$ functions," *Pacific J. Math.* **43** (1972), 175–188, MR 47 # 3670.

Krylov, V. I. "On functions regular in a half-plane," *Mat. Sb. (N.S.)* **6** (48) (1939), 95–138; *Amer. Math. Soc. Transl.* (2) **32** (1963), 37–81, MR 1,308.

Kuran, Ü. "A criterion of harmonic majorization in half-spaces," *Bull. London Math. Soc.* **3** (1971), 21–22, MR 47 # 3698.

Lax, P. D. "Translation invariant spaces," *Acta Math.* **101** (1959), 163–178, MR 21 # 4359; "Translation invariant spaces," in *Proc. Internat. Sympos. Linear Spaces* (Jerusalem, 1960), 299–306, Pergamon Press, New York (1961), MR 25 # 4345.

Leśniewicz, R. "On Hardy-Orlicz spaces. I," *Comment. Math. Prace Mat.* **15** (1971), 3–56, MR 54 # 13575; "On linear functionals in Hardy-Orlicz spaces. I-III," *Studia Math.* **46** (1973), 53–77 and 259–295; *ibid.* **47** (1973), 261–284, MR 58 # 23534a,b.

Levin, B. Ja. *Distribution of Zeros of Entire Functions*, revised ed., *Transl. Math. Monographs*, Vol. 5, Amer. Math. Soc., Providence, R. I. (1980), MR 81k:30011.

Levinson, N., and McKean, H. P., Jr. "Weighted trigonometrical approximation on R^1 with application to the germ field of a stationary Gaussian field," *Acta Math.* **112** (1964), 99–143, MR 29 # 414.

Levinson, N., et al. Articles dedicated to the memory of N. Wiener, *Bull. Amer. Math. Soc.* **72** (1966), 1–145, MR 32 # 1105,1106,1107.

Littlewood, J. E. "Mathematical notes (8); on functions subharmonic in a circle (II)," *Proc. London Math. Soc.* (2) **28** (1928), 383–394; *Lectures on the Theory of Functions*, Oxford University Press, London (1947), MR 6,261.

Loewner, C. (K. Löwner). "Über monotone Matrixfunktionen," *Math. Z.* **38** (1934), 177–216; "Some classes of functions defined by difference or differential inequalities," *Bull. Amer. Math. Soc.* **56** (1950), 308–319, MR 12,396.

Lowdenslager, D. "On factoring matrix valued functions," *Ann. of Math.* (2) **78** (1963), 450–454, MR 27 # 5094.

Macaev, V. I. "On the growth of entire functions that admit a certain estimate from below," *Dokl. Akad. Nauk SSSR* **132** (1960), 283–286; *Soviet Math. Dokl.* **1** (1960), 548–552, MR 23 # A328.

MacDuffee, C. C. *The Theory of Matrices*, Chelsea, New York (1946).

McEnnis, B. W. "Shifts on indefinite inner product spaces," *Pacific J. Math.* **81** (1979), 113–130; part II, *ibid.* **100** (1982), 177–183, MR 81c:47040 and 84b:47043.

McShane, E. J. *Integration*, Princeton University Press, Princeton, N. J. (1944), MR 6,43.

Marshall, D. E. "Blaschke products generate H^∞," *Bull. Amer. Math. Soc.* **82** (1976), 494–496, MR 53 # 5877.

Masani, P. R. "Sur les fonctions matricielles de la classe de Hardy H_2. I, II," *C. R. Acad. Sci. Paris* **249** (1959), 873–875; *ibid.* 906–907, MR 23 # A2235; "The prediction theory of multivariate stochastic processes. III," *Acta Math.* **104** (1960), 141–162, MR 22 # 12679; "Shift invariant spaces and prediction theory," *Acta Math.* **107** (1962),

275–290, MR 25#4344; "Recent trends in multivariate prediction theory," in *Multivariate Analysis*, P. R. Krishnaiah (ed.), Proc. Internat. Sympos. (Dayton, Ohio, 1965), 351–382, Academic Press, New York (1966), MR 35#5079.

Masani, P. R., and Wiener, N. "On bivariate stationary processes and the factorization of matrix-valued functions," *Theory Probab. Appl.* **4** (1959), 300–308, MR 23#A695.

Matveev, R. F. "Regularity of multidimensional stationary random processes with discrete time," *Dokl. Akad. Nauk SSSR* **126** (1959), 713–715, MR 22#6017.

Moore, B. "Outer Factorization for Vectorial Toeplitz Operators," Dissertation, University of Virginia (1969); (a) "The Szegö infimum," *Proc. Amer. Math. Soc.* **29** (1971), 55–62; *ibid.* **31** (1972), 638, MR 44#7355a,b; (b) "A factorable weight with zero Szegö infimum," *Proc. Amer. Math. Soc.* **35** (1972), 301–302, MR 48#935.

Moore, B., Rosenblum, M., and Rovnyak, J. "Toeplitz operators associated with isometries," *Proc. Amer. Math. Soc.* **49** (1975), 189–194, MR 51#11174.

Nagumo, M. "Über die gleichmässige Summierbarkeit und ihre Anwendung auf ein Variationsproblem," *Japan. J. Math.* **6** (1929), 173–182.

Netherton, J. "On Functions of Class $M(u,v)$," Dissertation, University of Virginia (1973).

von Neumann, J. (a) "Allgemeine Eigenwerttheorie Hermitscher Funktionaloperatoren," *Math. Ann.* **102** (1929), 49–131; Collected Works, Vol. II, no. 1; (b) "Zur Theorie der unbeschränkten Matrizen," *J.f. Math.* **161** (1929), 208–236; Collected Works, Vol. II, no. 3.

Nevanlinna, R. "Über beschränkte Funktionen, die in gegebenen Punkten vorgeschriebene Werte annehmen," *Ann. Acad. Sci. Fenn. Ser. A* **13**, no. 1 (1919), 71 pp.; (a) "Kriterien für die Randwerte beschränkter Funktionen," *Math. Z.* **13** (1922), 1–9; (b) "Asymptotische Entwicklungen beschränkter Funktionen und das Stieltjessche Momentenproblem," *Ann. Acad. Sci. Fenn. Ser. A* **18**, no. 5 (1922), 53 pp.; "Über beschränkte analytische Funktionen," *Ann. Acad. Sci. Fenn. Ser. A* **32**, no. 7 (1929), 75 pp.; "Beschränktartige Potentiale," *Math. Nachr.* **4** (1951), 489–501, MR 12,603.

Nielsen, O. A. *Direct Integral Theory*, Marcel Dekker, New York (1980).

Nikolaĭcuk, A. M., and Spĭtkovs'kiĭ, I. M. "On the Riemann boundary-value problem with hermitian matrix," *Dokl. Akad. Nauk SSSR* **221** (1975), 1280–1283; *Soviet Math. Dokl.* **16** (1975), 533–536, MR 53#8444.

Nikol'skiĭ, N. K. *Lectures on the Shift Operator*, "Nauka," Moscow (1980), MR 82i:47013.

Nudel'man, A. A. "On a new problem of moment type," *Dokl. Akad. Nauk SSSR* **233** (1977), 792–795; *Soviet Math. Dokl.* **18** (1977), 507–510, MR 57#10379; "On a generalization of classical interpolation problems," *Dokl. Akad. Nauk SSSR* **256** (1981), 790–793; *Soviet Math. Dokl.* **23** (1981), 125–128, MR 82f:30033.

Page, L. B. "Applications of the Sz.-Nagy and Foiaş lifting theorem," *Indiana Univ. Math. J.* **20** (1970), 135–145, MR 41#6003; "The Szegö and Beurling theorems and operator factorizations," in *Spectral Theory*, Banach Center Publications 8, 355–360, PWN-Polish Scientific Publishers, Warsaw (1982).

Paley, R. E. A. C., and Wiener, N. *Fourier Transforms in the Complex Domain*, Amer. Math. Soc. Coll. Publ., Vol. 19, New York (1934).

Parreau, M. "Sur les moyennes des fonctions harmoniques et analytiques et la classification des surfaces de Riemann," *Ann. Inst. Fourier (Grenoble)* **3** (1952), 103–197, MR 14,263.

Pfluger, A. "Über konforme Abbildungen des Einheitskreises," *Ann. Acad. Sci. Fenn. Ser. A I Math.* **7** (1982), 73–79, MR 84e:01045.

Pick, G. "Über die Beschränkungen analytischer Funktionen, welche durch vorgegebene

Funktionswerte bewirkt werden," *Math. Ann.* **77** (1916), 7–23; "Über die Be-schränkungen analytischer Funktionen durch vorgegebene Funktionswerte," *Math. Ann.* **78** (1918), 270–275; "Über beschränkte Funktionen mit vorgeschriebenen Wertzuordnungen," *Ann. Acad. Sci. Fenn. Ser. A* **15**, no. 3 (1920), 17 pp.

Pincus, J. D. "Commutators, generalized eigenfunction expansions and singular integral operators," *Trans. Amer. Math. Soc.* **121** (1966), 358–377, MR 32 # 6228; "Commu-tators and systems of singular integral equations. I," *Acta Math.* **121** (1968), 219–249, MR 39 # 2026.

Pincus, J. D., and Rovnyak, J. "A spectral theory for some unbounded self-adjoint singular integral operators," *Amer. J. Math.* **91** (1969), 619–636, MR 40 # 3373.

Pitt, L. D. "Weighted L^p closure theorems for spaces of entire functions," *Israel J. Math.* **24** (1976), 94–118, MR 57 # 17239; "A general approach to approximation problems of the Bernstein type," *Adv. in Math.* **49** (1983), 264–299.

Popov, V. M. *Hyperstability of Control Systems*, Springer-Verlag, New York (1973), MR 52 # 8588.

Potapov, V. P. "The multiplicative structure of *J*-contractive matrix functions," *Trudy Moskov. Mat. Obshch.* **4** (1955), 125–236; *Amer. Math. Soc. Transl.* (2) **15** (1960), 131–243, MR 17,958.

Putnam, C. R. "On Toeplitz matrices, absolute continuity, and unitary equivalence," *Pacific J. Math.* **9** (1959), 837–846, MR 22 # 183; *Commutation Properties of Hilbert Space Operators*, Springer-Verlag, New York (1967), MR 36 # 707.

Radjavi, H., and Rosenthal, P. *Invariant Subspaces*, Springer-Verlag, New York (1973), MR 51 # 3924.

Radó, T. *Subharmonic Functions*, Chelsea, New York (1949).

Ran, A. C. M. *Semidefinite Invariant Subspaces*, Brügemann B. V., Den Burg-Texel, Netherlands (1984).

Riesz, F., and Sz.-Nagy, B. *Functional Analysis*, Frederick Ungar, New York (1955), MR 17,175.

Rivlin, T. J. *The Chebyshev Polynomials*, Wiley, New York (1974), MR 56 # 9142.

Robinson, E. A. (a) "Extremal representation of stationary stochastic processes," *Ark. Mat.* **4** (1962), 379–384, MR 25 # 4576; (b) *Random Wavelets and Cybernetic Systems*, Charles Griffin and Company Limited, London (1962), MR 26 # 7457.

Rochberg, R. "The equation $(I - S)g = f$ for shift operators in Hilbert space," *Proc. Amer. Math. Soc.* **19** (1968), 123–129, MR 36 # 5722.

Rosenblatt, M. "A multi-dimensional prediction problem," *Ark. Mat.* **3** (1958), 407–424, MR 19,1098.

Rosenblum, M. "On the Hilbert matrix. I,II," *Proc. Amer. Math. Soc.* **9** (1958), 137–140 and 581–585, MR 20 # 1139,6038; "The absolute continuity of Toeplitz's matrices," *Pacific J. Math.* **10** (1960), 987–996, MR 22 # 4952; "Self-adjoint Toeplitz operators and associated orthonormal functions," *Proc. Amer. Math. Soc.* **13** (1962), 590–595, MR 25 # 1449; "A concrete spectral theory for self-adjoint Toeplitz operators," *Amer. J. Math.* **87** (1965), 709–718, MR 31 # 6127; "A spectral theory for self-adjoint singular integral operators," *Amer. J. Math.* **88** (1966), 314–328, MR 33 # 6453; "Vectorial Toeplitz operators and the Fejér-Riesz theorem," *J. Math. Anal. Appl.* **23** (1968), 139–147, MR 37 # 3378; "A corona theorem for countably many functions," *Integral Equations and Operator Theory* **3** (1980), 125–137, MR 81e:46034.

Rosenblum, M., and Rovnyak, J. "Factorization of operator valued entire functions," *Bull. Amer. Math. Soc.* **75** (1969), 1343–1346, MR 40 # 3349; "Factorization of operator valued entire functions," *Indiana Univ. Math. J*. **20** (1970), 157–173, MR 41 # 6005; "The factorization problem for nonnegative operator valued functions," *Bull. Amer. Math. Soc*. **77** (1971), 287–318, MR 42 # 8315; "Two theorems on finite Hilbert transforms," *J. Math. Anal. Appl*. **48** (1974), 708–720, MR 51 # 1259; "Restrictions of analytic functions. I-III," *Proc. Amer. Math. Soc*. **48** (1975), 113–119; *ibid*. **51** (1975), 335–343; *ibid*. **52** (1975), 222–226, MR 53 # 3764a,b,c; "Change of variables formulas with Cayley inner functions," in *Topics in Functional Analysis* (essays dedicated to M. G. Kreĭn on the occasion of his 70th birthday), 283–320, *Adv. in Math. Suppl. Stud*., Vol. 3, Academic Press, New York (1978), MR 81d:30053; "An operator-theoretic approach to theorems of the Pick-Nevanlinna and Loewner types. I, II," *Integral Equations and Operator Theory* **3** (1980), 408–436; *ibid*. **5** (1982), 870–887, MR 82a:47016 and 84b:47020.

Rota, G. C. "Note on the invariant subspaces of linear operators," *Rend. Circ. Mat. Palermo* (2) **8** (1959), 182–184, MR 22 # 2897; "On models for linear operators," *Comm. Pure Appl. Math*. **13** (1960), 469–472, MR 22 # 2898.

Rovnyak, J. "Ideals of square summable power series," *Proc. Amer. Math. Soc*. **13** (1962), 360–365; part II, *ibid*. **16** (1965), 209–212, MR 25 # 2455 and 30 # 3358; "Some Hilbert Spaces of Analytic Functions," Dissertation, Yale University (1963); "On the theory of unbounded Toeplitz operators," *Pacific J. Math*. **31** (1969), 481–496, MR 40 # 6288; "The absolutely continuous component of a selfadjoint Toeplitz operator," *Indiana Univ. Math. J*. **21** (1972), 751–757, MR 44 # 7356; "A converse to von Neumann's inequality," *Proc. Amer. Math. Soc*. **84** (1982), 370–372, MR 82k:47021.

Rozanov, Yu. A. "Spectral theory of multi-dimensional stationary random processes with discrete time," *Uspehi Mat. Nauk (N.S.)* **13** (1958), 93–142; *Selected Translations in Mathematical Statistics and Probability*, Vol. 1, 253–306, Amer. Math. Soc., Providence, R. I. (1961), MR 22 # 7168; "Spectral properties of multivariate stationary processes and boundary properties of analytic matrices," *Theory Probab. Appl*. **5** (1960), 362–376, MR 24 # A2432; *Stationary Random Processes*, Moscow (1963); Holden-Day, San Francisco (1967), MR 35 # 4985; *Innovation Processes*, Winston, Washington (1977), MR 56 # 3932.

Rudin, W. "Analytic functions of class H_p," *Trans. Amer. Math. Soc*. **78** (1955), 46–66, MR16,810; *Real and Complex Analysis*, McGraw-Hill, New York (1966), MR 35 # 1420; *Function Theory in Polydisks*, W. A. Benjamin, New York (1969), MR 41 # 501.

Saal, R. "Activities on network theory and circuit design in Europe," *IEEE Trans. Circuits and Systems* **CAS-31** (1984), 124–133.

Sarason, D. "Generalized interpolation in H^∞," *Trans. Amer. Math. Soc*. **127** (1967), 179–203, MR 34 # 8193.

Schur, I. "Über Potenzreihen, die im Innern des Einheitskreises beschränkt sind. I,II," *J. Reine Angew. Math*. **147** (1917), 205–232; *ibid*. **148** (1918), 122–145; *Gesammelte Abhandlungen*, Vol. II, nos. 29, 30.

Shapiro, H. S. "Generalized analytic continuation," *Symposia on Theoretical Physics and Mathematics*, Vol. 8 (Symposium, Madras, 1967), 151–163, Plenum, New York (1968), MR 39 # 2953.

Šmul'jan, Ju. L. "Monotone operator functions on a set consisting of an interval and a point," *Ukrain. Mat. Zh*. **17** (1965), 130–136; *Amer. Math. Soc. Transl*. (2) **67** (1968), 25–32, MR 32 # 6239.

Solomentsev, E. "On some classes of subharmonic functions," *Bull. Acad. Sci. URSS Ser. Math. Nr*. **516** (1938), 571–582.

Sparr, G. "A new proof of Löwner's theorem on monotone matrix functions," *Math. Scand.* **47** (1980), 266–274, MR 82e:47021.

Stone, M. H. *Linear Transformations in Hilbert Space*, Amer. Math. Soc. Coll. Publ., Vol. 15, New York (1932).

Stray, A. "A formula by V. M. Adamjan, D. Z. Arov and M. G. Kreĭn," *Proc. Amer. Math. Soc.* **83** (1981), 337–340, MR 83c:30029; "Minimal interpolation by Blaschke products," preprint (1984).

Suzuki, N. "The finite Hilbert transform on $L_2(0,\pi)$ is a shift," *Proc. Japan Acad.* **52** (1976), 544–547, MR 55 # 3881.

Szegö, G. "Über die Randwerte analytischer Funktionen," *Math. Ann.* **84** (1921), 232–244; *Orthogonal Polynomials*, 4th ed., Amer. Math. Soc. Coll. Publ., Vol. 23, Providence, R. I. (1975), MR 51 # 8724.

Sz.-Nagy, B. "Remarks to the preceding paper of A. Korányi," *Acta Sci. Math. (Szeged)* **17** (1956), 71–75, MR 18,588.

Sz.-Nagy, B., and Foiaş, C. "Sur les contractions de l'espace de Hilbert. VIII Fonctions caractéristiques. Modèles fonctionnels," *Acta Sci. Math. (Szeged)* **25** (1964), 38–71; "– – –. XII Fonctions intérieures, admettant des facteurs extérieurs," *ibid.* **27** (1966), 27–33, MR 30 # 2348 and 33 # 6414; (a) "Commutants de certains opérateurs," *Acta Sci. Math. (Szeged)* **29** (1968), 1–17, MR 39 # 3346; (b) "Dilatation des commutants d'opérateurs," *C. R. Acad. Sci. Paris Sér. A* **266** (1968), 493–495, MR 38 # 5049; *Harmonic Analysis of Operators on Hilbert Space*, North Holland, New York (1970), MR 43 # 947.

Sz.-Nagy, B., and Korányi, A. "Relations d'un problème de Nevanlinna et Pick avec la théorie des opérateurs de l'espace hilbertien," *Acta Math. Acad. Sci. Hungar.* **7** (1956), 295–303, MR 19,296; "Operatortheoretische Behandlung und Verall-gemeinerung eines Problemkreises in der komplexen Funktionentheorie," *Acta Math.* **100** (1958), 171–202, MR 24 # A437.

Takahashi, K. "The factorization in the commutant of a unitary operator," *Hokkaido Math. J.* **8** (1979), 253–259, MR 81b:47024.

Titchmarsh, E. C. *Theory of Fourier Integrals*, 2nd ed., Oxford University Press, London (1948).

Toeplitz, O. "Zur Transformationen der Scharen bilinearer Formen von unendlichvielen Veränderlichen," *Nachr. Akad. Wiss. Göttingen: Math. Phys. Kl.* (1907), 110–115; "Zur Theorie der quadratischen Formen von unendlichvielen Variablen," *Nachr. Akad. Wiss. Göttingen: Math. Phys. Kl.* (1910), 489–506; (a) "Zur Theorie der quadratischen und bilinearen Formen von unendlichvielen Veränderlichen," *Math. Ann.* **70** (1911), 351–376; (b) "Über die Fourier'sche Entwicklung positiver Funktionen," *Rend. Circ. Mat. Palermo* **32** (1911), 191–192.

Tolsted, E. "Limiting values of subharmonic functions," *Proc. Amer. Math. Soc.* **1** (1950), 636–647, MR 12,609; "Non-tangential limits of subharmonic functions," *Proc. London Math. Soc.* (3) **7** (1957), 321–333, MR 20 # 1110.

Tricomi, F. G. *Integral Equations*, Interscience, New York (1957), MR 20 # 1177.

Tsuji, M. *Potential Theory in Modern Function Theory*, Chelsea, New York (1959), MR 22 # 5712.

Tumarkin, G. C. "Description of a class of functions admitting approximation by fractions with preassigned poles," *Izv. Akad. Nauk Armjan. SSR Ser. Mat.* **1** (1966), 89–105, MR 34 # 6123.

de la Vallée Poussin, Ch. J. "Sur l'intégral de Lebesgue," *Trans. Amer. Math. Soc.* **16** (1915), 435–501.

Vasudeva, H. "On monotone matrix functions of two variables," *Trans. Amer. Math. Soc.* **176** (1973), 305–318, MR 47 #2408.

Walsh, J. L. *Interpolation and Approximation*, Amer. Math. Soc. Coll. Publ., Vol. 20, New York (1935).

Weiss, G. "A note on Orlicz spaces," *Portugal. Math.* **15** (1956), 35–47, MR 18,586.

Widom, H. "Toeplitz matrices," in *Studies in Real and Complex Analysis*, Vol. 3, I. I. Hirschman, Jr. (ed.), 179–209, Mathematical Association of America, Prentice-Hall, Englewood Cliffs, N. J. (1965), MR 32 #1080.

Wiener, N. *Extrapolation, Interpolation, and Smoothing of Time Series*, Wiley, New York (1949) (originally a classified report dated February, 1942); paperback, M.I.T. Press, Cambridge (1964), MR 11,118; "On the factorization of matrices," *Comment. Math. Helv.* **29** (1955), 97–111, MR 16,921.

Wiener, N., and Akutowicz, E. J. "A factorization of positive hermitian matrices," *J. Math. Mech.* **8** (1959), 111–120, MR 21 #2158.

Wiener, N., and Masani, P. R. "The prediction theory of multivariate stochastic processes. I, II," *Acta Math.* **98** (1957), 111–150; *ibid.* **99** (1958), 93–137, MR 20 #4323,4325.

Wigner, E. P., and von Neumann, J. "Significance of Loewner's theorem in the quantum theory of collisions," *Ann. of Math.* (2) **59** (1954), 418–433, MR 16,25.

Williams, J. P. "Minimal spectral sets of compact operators," *Acta Sci. Math. (Szeged)* **28** (1967), 93–106, MR 36 #725.

Wishard, A. "Functions of bounded type," *Duke Math. J.* **9** (1942), 663–676, MR 5,115.

Wold, H. *A Study in the Analysis of Stationary Time Series*, Almqvist and Wiksell, Stockholm (1938); 2nd ed. (1954), MR 15,811.

Young, N. J. "The Nevanlinna-Pick problem for matrix-valued functions," preprint, 1984.

Zasuhin, V. N. "On the theory of multidimensional stationary random processes," *Dokl. Akad. Nauk SSSR* **33** (1941), 435–437, MR 5,102.

Ziegler, H. J. W. *Vector Valued Nevanlinna Theory*, Pitman, Boston (1982), MR 84d:30057.

Zygmund, A. *Trigonometric Series I, II*, 2nd ed., Cambridge University Press, London (1968), MR 38 #4882.

Author Index

Subject Index